TUMORS and CANCERS

ENDOCRINE GLANDS—BLOOD—MARROW—LYMPH

POCKET GUIDES TO
BIOMEDICAL SCIENCES

https://www.crcpress.com/Pocket-Guides-to-Biomedical-Sciences/book-series/CRCPOCGUITOB

The ***Pocket Guides to Biomedical Sciences*** series is designed to provide a concise, state-of-the-art, and authoritative coverage on topics that are of interest to undergraduate and graduate students of biomedical majors, health professionals with limited time to conduct their own searches, and the general public who are seeking for reliable, trustworthy information in biomedical fields.

TUMORS
and
CANCERS

ENDOCRINE GLANDS—BLOOD—MARROW—LYMPH

Dongyou Liu

CRC Press is an imprint of the
Taylor & Francis Group, an **informa** business

CRC Press
Taylor & Francis Group
6000 Broken Sound Parkway NW, Suite 300
Boca Raton, FL 33487-2742

Printed on acid-free paper

International Standard Book Number-13: 978-1-4987-2975-8 (Paperback)
978-1-1383-0087-3 (Hardback)

Library of Congress Cataloging-in-Publication Data

Names: Liu, Dongyou, author.
Title: Tumors and cancers : endocrine glands - blood - marrow - lymph / Dongyou Liu.
Description: Boca Raton : Taylor & Francis, a CRC title, part of the Taylor & Francis imprint, a member of the Taylor & Francis Group, the academic division of T&F Informa plc, 2018. | Series: Pocket guides to biomedical sciences | Includes bibliographical references and index.
Identifiers: LCCN 2017029330| ISBN 9781498729758 (paperback : alk. paper) | ISBN 9781138300873 (hardback : alk. paper)
Subjects: LCSH: Endocrine glands--Tumors. | Endocrine glands--Cancer.
Classification: LCC RC280.E55 L58 2018 | DDC 616.99/44--dc23
LC record available at https://lccn.loc.gov/2017029330

Visit the Taylor & Francis Web site at
http://www.taylorandfrancis.com

and the CRC Press Web site at
http://www.crcpress.com

Contents

Series Preface

Dramatic breakthroughs and nonstop discoveries have rendered biomedicine increasingly relevant to everyday life. Keeping pace with all these advances is a daunting task, even for active researchers. There is an obvious demand for succinct reviews and synthetic summaries of biomedical topics for graduate students, undergraduates, faculty, biomedical researchers, medical professionals, science policy makers, and the general public.

Recognizing this pressing need, CRC Press has established the Pocket Guides to Biomedical Science series, with the main goal to provide state-of-the-art, authoritative reviews of far-ranging subjects in short, readable formats intended for a broad audience. Volumes in the series will address and integrate the principles and concepts of the natural sciences and liberal arts, especially those relating to biomedicine and human well-being. Future volumes will come from biochemistry, bioethics, cell biology, genetics, immunology, microbiology, molecular biology, neuroscience, oncology, parasitology, pathology, and virology, as well as from other related disciplines.

Forming part of the four volumes devoted to human tumors and cancers, the current volume focuses on the endocrine, hematopoietic, and lymphoreticular systems. Characterized by uncontrolled growth of abnormal cells that often extend beyond their usual boundaries and disrupt the normal functions of affected organs, tumors and cancers are insidious diseases with serious consequences. Relative to our ongoing research and development efforts, our understanding of tumors and cancers remains rudimentary, and arsenals at our disposal against these increasingly prevalent diseases are severely limited. The goal of this volume is the same as the goal for the series—to simplify, summarize, and synthesize a complex topic so that readers can reach to the core of the matter without the necessity of carrying out their own time-consuming literature searches.

We welcome suggestions and recommendations from readers and members of the biomedical community for future topics in the series as well as experts to serve as potential volume authors/editors.

Dongyou Liu
Sydney, Australia

Contributors

Francesco Facciolo, MD
Thoracic Surgery Unit
Regina Elena National Cancer Institute
Rome, Italy

Rina Kansal, MD
Blood Center of Wisconsin
Milwaukee, Wisconsin

Jian Yi Li, MD, PhD
Department of Pathology and Laboratory Medicine
North Shore University Hospital and Long Island Jewish Medical Center
Northwell Health
Hofstra Northwell School of Medicine
Lake Success, New York

Dongyou Liu, PhD
RCPA Quality Assurance Programs
Sydney, New South Wales, Australia

Mirella Marino, MD
Department of Pathology
Regina Elena National Cancer Institute
Rome, Italy

Estella Matutes, MD, PhD, FRCPath
Hematopathology Unit
Hospital Clinic of Barcelona
Barcelona, Spain

Maria Teresa Ramieri, MD
Department of Pathology
"F. Spaziani" Hospital
ASL Frosinone
Frosinone, Italy

1 Introductory Remarks

1.1 Preamble

Tumor or cancer (these terms, along with *neoplasm* and *lesion*, are used interchangeably in colloquial language and publications; see Glossary) is an insidious disease that results from an uncontrolled growth of abnormal cells in part(s) of the body. Tumor/cancer has acquired a notorious reputation due not only to its ability to exploit host cellular machineries for its own advantages but also to its potential to cause human misery.

With a rapidly aging world population, widespread oncogenic viruses, and constant environmental pollution and destruction, tumor/cancer is poised to exert an increasingly severe toll on human health and well-being. There is a burgeoning interest from health professionals and the general public in learning about tumor/cancer mechanisms, clinical features, diagnosis, treatment, and prognosis. The following pages in the current volume as well as those in the sister volumes represent a concerted effort to satisfy this critical need.

1.2 Tumor mechanisms

The human body is composed of various types of cells that grow, divide, and die in an orderly fashion (a process known as apoptosis). However, when some cells in the body change their growth patterns and fail to undergo apoptosis, they often produce a solid tumor or sometimes a nonsolid tumor (e.g., leukemia). A tumor is considered benign if it grows but does not spread beyond the immediate area in which it arises. While most benign tumors are not life-threatening, those found in the vital organs (e.g., the brain) can be deadly. In addition, some benign tumors are precancerous, with the propensity to become cancerous if left untreated. On the contrary, a tumor is considered malignant and cancerous if it grows continuously and spreads to surrounding areas and other parts of the body through the blood or lymph system.

A tumor located in its original (primary) site is known as a "primary tumor." A tumor that spreads from its original (primary) site via the neighboring tissue, bloodstream, or lymphatic system to another site in the body is called a "metastatic tumor/cancer" (or secondary tumor/cancer). A metastatic cancer has the same name and the same type of cancer cells as a primary cancer. For instance, a metastatic cancer in the brain that originates from breast cancer is known as metastatic breast cancer—not brain cancer.

Typically, tumors/cancers form in tissues after the cells undergo genetic mutations that lead to sequential changes known as hyperplasia, metaplasia, dysplasia, neoplasia, and anaplasia (see Glossary). Factors contributing to genetic mutations in the cells may be chemical (e.g., cigarette smoking, asbestos, paint, dye, bitumen, mineral oil, nickel, arsenic, aflatoxin, and wood dust), physical (e.g., sun, heat, radiation, and chronic trauma), viral (e.g., EBV, HBV, HPV, and HTLV-1), immunological (e.g., AIDS and transplantation), endocrine (e.g., excessive endogenous or exogenous hormones), or hereditary (e.g., familial inherited or non-inherited disorders).

In essence, tumorigenesis is a cumulative process that demonstrates several notable hallmarks, including (i) sustaining proliferative signaling, (ii) activating local invasion and metastasis, (iii) resisting apoptosis and enabling replicative immortality, (iv) inducing angiogenesis and inflammation, (v) evading immune destruction, (vi) deregulating cellular energetics, and (vii) genome instability and mutation.

1.3 Tumor classification, grading, and staging

A tumor/cancer is usually named for the organs or tissues where it starts (e.g., brain cancer, breast cancer, lung cancer, lymphoma, and skin cancer). Depending on the types of tissue involved, tumors/cancers are grouped into a number of broad categories: (i) carcinoma (involving epithelium), (ii) sarcoma (involving soft tissue), (iii) leukemia (involving blood-forming tissue), (iv) lymphoma (involving lymphocytes), (v) myeloma (involving plasma cells), (vi) melanoma (involving melanocytes), (vii) central nervous system cancers (involving brain or spinal cord), (viii) germ cell tumors (involving cells that give rise to sperm or eggs), (ix) neuroendocrine tumors (involving hormone-releasing cells), and (x) carcinoid tumors (a variant of neuroendocrine tumors found mainly in the intestinal tract).

Primary tumors of the endocrine system include those affecting the adrenal, pituitary, parathyroid, and thyroid glands as well as the endocrine pancreas. In addition, inherited and non-inherited tumor syndromes (e.g., Carney complex;

Li–Fraumeni syndrome; Mahvash disease; McCune–Albright syndrome; multiple endocrine neoplasia types 1, 2, and 4; von Hippel–Lindau syndrome [VHL]; neurofibromatosis type 1; hyperparathyroidism-jaw tumor syndrome; familial paraganglioma–pheochromocytoma syndrome; and familial non-medullary thyroid cancer) are known to predispose the endocrine glands and other organs to a range of neoplasms [1].

Tumors of the hematopoietic and lymphoreticular systems consist of two main categories: (i) myeloid neoplasms and acute leukemias (which account for about 35% of all hematopoietic and lymphoreticular neoplasms) and (ii) mature lymphoid, histiocytic, and dendritic neoplasms (which include myeloma and represent 65% of all hematopoietic and lymphoreticular neoplasms) [2–5].

Myeloid neoplasms and acute leukemias are further separated into 10 groups: (i) myeloproliferative neoplasms (MPN), (ii) mastocytosis, (iii) myeloid/lymphoid neoplasms with eosinophilia and rearrangements, (iv) myelodysplastic/myeloproliferative neoplasms (MDS/MPN), (v) myelodysplastic syndrome (MDS), (vi) acute myeloid leukemia (AML), (vii) blastic plasmacytoid dendritic cell neoplasms (BPDCN) (viii) acute leukemia of ambiguous lineage (ALAL), (ix) B-lymphoblastic leukemia/lymphoma, and (x) T-lymphoblastic leukemia/lymphoma [4].

Similarly, mature lymphoid, histiocytic, and dendritic neoplasms are subdivided into five groups: (i) mature B-cell neoplasms, (ii) mature T and NK neoplasms, (iii) Hodgkin lymphoma, (iv) posttransplant lymphoproliferative disorders (PTLD), and (v) histiocytic and dendritic cell neoplasms. Of these, mature B-cell neoplasms and mature T and NK neoplasms are commonly referred to as non-Hodgkin lymphoma (NHL) [5].

Under the auspices of the World Health Organization (WHO), the International Classification of Diseases for Oncology, third edition (ICD-O-3) [6], has designed a five-digit system for classifying tumors, with the first four digits being the morphology code and the fifth digit being the behavior code [6]. The fifth digit behavior codes for neoplasms range from 0 (benign), 1 (benign or malignant), 2 (carcinoma *in situ*), 3 (malignant, primary site), 6 (malignant, metastatic site) to 9 (malignant, primary or metastatic site). For example, chondroma has an IDC-O-3 code of 9220/0 and is considered a benign bone tumor; multiple chondromatosis (a subtype of chondroma) has an IDC-O-3 code of 9220/1 and is an intermediate grade bone tumor with the potential for malignant transformation; and central chondrosarcoma has an IDC-O-3 code of 9220/3 and is considered a malignant bone tumor [6].

To further delineate tumors/cancers and assist in their treatment and prognosis, the pathological stages of solid tumors are often determined by using the TNM system (see Glossary) of the American Joint Commission on Cancer (AJCC), which incorporates the size and extent of the primary tumor (TX, T0, T1, T2, T3, T4), the number of nearby lymph nodes involved (NX, N0, N1, N2, N3), and the presence of distant metastasis (MX, M0, M1) [7]. Therefore, under the TNM system, the pathological stage of a given tumor or cancer may be referred to as T1N0MX or T3N1M0 (with numbers after each letter providing further details about the tumor or cancer). However, a much more simplified staging system (0, I, II, III, IV) is adopted clinically to describe the stages of solid tumors (see "stage" in Glossary) [7].

Another staging system that is more often used by cancer registries than by doctors divides tumors/cancers into five categories: (i) *in situ* (abnormal cells are present but have not spread to nearby tissue); (ii) localized (cancer is limited to the place where it started, with no sign that it has spread); (iii) regional (cancer has spread to nearby lymph nodes, tissues, or organs); (iv) distant (cancer has spread to distant parts of the body); and (v) unknown (there is not enough information to determine the stage).

1.4 Tumor diagnosis

As most tumors/cancers tend to induce nonspecific, noncharacteristic clinical signs, a variety of procedures and tests are utilized during a diagnostic workup. These involve a medical history review of the patient and their relatives (for clues to potential risk factors that enhance cancer development), a complete physiological examination (for lumps and other abnormalities), imaging techniques (e.g., ultrasound, CT, MRI, and PET; see Glossary), biochemical and immunological tests (for altered substance or cell levels in blood, bone marrow, cerebrospinal fluid, urine, and tissue), histological evaluation of biopsy and tissue (using hematoxylin and eosin [H&E] and immunohistochemical [IHC] stains; see Glossary), and molecular analyses (e.g., FISH and PCR; see Glossary).

1.5 Tumor treatment and prognosis

Standard cancer treatments consist of surgery (for removal of tumors and relieving symptoms associated with tumors), radiotherapy (also called radiation therapy or X-ray therapy; delivered externally through the skin or internally [brachytherapy] for destruction of cancer cells or impeding their growth), chemotherapy (for inhibiting the growth of cancer cells, suppressing the body's hormone production, blocking the effect of the hormone on cancer cells, etc.—usually via the bloodstream or oral ingestion), and complementary therapies (for enhancing patients' quality of life and improving their well-being). Depending on the circumstances, surgery may be used in combination with radiotherapy and/or chemotherapy to ensure that any cancer cells remaining in the body are eliminated.

The outcomes of tumor/cancer treatments include (i) cure (no traces of cancer remain after treatment, and the cancer will never come back), (ii) remission (signs and symptoms of cancer are reduced; in a complete remission, all signs and symptoms of cancer disappearing for 5 years or more suggest a cure), or (iii) recurrence (a benign or cancerous tumor comes back after surgical removal and adjunctive therapy).

Prognosis (or chance of recovery) for a given tumor/cancer is usually dependent on the location, type, and grade of the tumor, patient's age and health status, etc. Regardless of tumor/cancer types, patients with lower grade lesions generally have a better prognosis than those with higher grade lesions.

1.6 Future perspective

Tumor/cancer is a biologically complex disease that is expected to surpass heart disease to become the leading cause of human death throughout the world in the coming decades. Despite extensive past research and development efforts, tumor/cancer remains poorly understood, and effective cures remain largely elusive.

The completion of the Human Genome Project in 2003 and the establishment of The Cancer Genome Atlas (TCGA) in 2005 have offered promises for better understanding of the genetic basis of human tumors/cancers and have opened new avenues for developing novel diagnostic techniques and effective therapeutic measures.

Nonetheless, a multitude of factors pose continuing challenges for the ultimate conquest of tumors/cancers. These include the inherent biological complexity and heterogeneity of tumors/cancers, contribution of various genetic and environmental risk factors, absence of suitable models for human tumors/cancers, and difficulty in identifying therapeutic compounds that kill/inhibit cancer cells only and not normal cells. Further efforts are necessary to help overcome these obstacles and to enhance the well-being of cancer sufferers.

Acknowledgments

Credit is due to a group of international oncologists/clinicians whose expert contributions have greatly enriched this volume.

References

1. DeLellis RA. (ed). *Pathology and genetics of tumours of endocrine organs. International Agency for Research on Cancer*; World Health Organization; International Academy of Pathology; International Association for the Study of Lung Cancer. Lyon: IARC Press, 2004.
2. Jaffe ES, Harris NL, Stein H, Vardiman JW. *Pathology and genetics of tumours of haematopoietic and lymphoid tissues*; World Health Organization; Lyon: IARC Press; Oxford University Press (distributor), 2001.
3. Swerdlow SH, Campo E, Harris NL, et al. (eds). *WHO classification of tumours of haematopoietic and lymphoid tissues*. International Agency for Research on Cancer; World Health Organization. Lyon, France: IARC, 2008.
4. Arber DA, Orazi A, Hasserjian R, et al. The 2016 revision to the World Health Organization classification of myeloid neoplasms and acute leukemia. *Blood*. 2016; 127(20):2391–405.
5. Swerdlow SH, Campo E, Pileri SA, et al. The 2016 revision of the World Health Organization classification of lymphoid neoplasms. *Blood* 2016; 127(20):2375.
6. Fritz A, Percy C, Jack A, et al. *International classification of diseases for oncology* (3rd ed). Geneva: World Health Organization, 2000.
7. Edge SB, Byrd DR, Compton CC, Fritz AG, Greene FL, Trotti A, (eds). *AJCC cancer staging manual* (7th ed). New York, NY: Springer, 2010.

SECTION I

Endocrine System

2 Adrenal Tumors

2.1 Definition

Tumors of the adrenal glands include neoplasms affecting the adrenal cortex (i.e., adrenocortical tumors) and those affecting the adrenal medulla (i.e., adrenal medullary tumors).

Adrenocortical tumors are essentially epithelial tumors that can be divided into adrenal cortical adenoma (ACA) and adrenal cortical carcinoma (ACC). Accounting for a majority (>95%) of adrenal neoplasms, ACA (also known as adrenal adenoma) is benign, often nonfunctional, and clinically irrelevant. ACC is a rare, malignant neoplasm (1% of adrenal tumors) with a dismal prognosis [1,2].

Adrenal medullary tumors are relatively uncommon and can be broadly classified into sympathetic system tumors (e.g., neuroblastoma, ganglioneuroblastoma, and ganglioneuroma) and adult neuroendocrine tumors (e.g., pheochromocytoma and adrenal medullary hyperplasia), in addition to several recently proposed tumors (e.g., sustentaculoma and corticomedullary mixed tumor [i.e., mixed ACA-pheochromocytoma]).

2.2 Biology

Located above and slightly medial to the kidneys, the adrenal glands (also known as the suprarenal glands) are a pair of small organs each approximately 5 cm by 3 cm by 1 cm in dimension and 4–6 g in weight. In adults, the left gland is crescentic and the right gland is pyramidal.

Structurally, the adrenal glands are covered by a fibrous capsule and comprise two main parts: the outer cortex and inner medulla. The adrenal cortex (derived from mesothelium, 2 cm thick) is bright yellow in color (due to the presence of lipid) and is further separated into three zones: the zona glomerulosa, zona fasciculata, and zona reticularis.

The zona glomerulosa (outer layer, 15% of cortex volume) is composed of small clusters and short trabeculae of relatively small, well-defined cells; it is involved in the production of mineralocorticoids such as aldosterone for regulation of blood pressure and electrolyte balance. The zona fasciculata (middle layer, 80% of cortex volume) is composed of large cells with distinct membranes arranged in cords two cells wide and cytoplasm containing small lipid vacuoles; it is involved in the production of glucocorticoids such as 11-deoxycorticosterone, corticosterone, and cortisol, for regulation of metabolism and immune system suppression. The zona reticularis (inner layer, 5% of cortex volume) is composed of haphazardly arranged small cells with granular and eosinophilic cytoplasm; it is involved in the production of androgens such as dehydroepiandrosterone (DHEA), DHEA sulfate (DHEA-S), and androstenedione (testosterone precursor) for sexual development and functions.

The adrenal medulla is ellipsoid in shape, is gray-tan in color, and makes up <10% of gland volume (1% in neonates). Derived from neural crest and mostly found within the head of the gland, the adrenal medulla is composed of neural crest cells (also called chromaffin cells, pheochromocytes, or medullary cells, which are large polygonal cells arranged in small nests and cords separated by prominent vasculature, with poorly outlined borders and abundant granular and basophilic cytoplasm) and sustentacular cells (spindle-shaped supporting cells at the periphery of nests of chromaffin cells). The chromaffin cells are the key supplier of catecholamines, including epinephrine (adrenaline, 80%) and norepinephrine (noradrenaline, 20%), for regulation of blood pressure and heart rate.

Glucocorticoids are regulated by the hypothalamus–pituitary–adrenal (HPA) axis through the adrenocorticotropic hormone (ACTH, released by the anterior pituitary) and the corticotropin-releasing hormone (CRH, released by neurons of the hypothalamus). Mineralocorticoid secretion is influenced by the renin–angiotensin–aldosterone system (RAAS), the concentration of potassium, and the concentration of ACTH.

ACA is a benign neoplasm resulting from neoplastic proliferation of adrenal cortical cells (of all three zones but more commonly zona fasciculata). Often discovered accidentally (so-called incidentaloma), ACA may appear as black (pigmented) adenoma (diffusely pigmented, brown–black neoplasm due to lipofuscin) and may or may not be functional.

ACC may occur sporadically or progress stepwise from a low-grade to high-grade carcinoma. There is a well-known association between ACC and various genetic syndromes [1,2].

Pheochromocytoma evolves from chromaffin cells in the adrenal medulla and is considered a hormonally silent incidentaloma.

2.3 Epidemiology

Adrenal tumors affect 3%–10% of the human population and are predominated by benign nonfunctional ACA (constituting about 80% of adrenal tumors) along with some functional ACA (15% of adrenal tumors). ACA is found in <1% of people under 30 years of age and in 7% of people >70 years of age and shows a female bias.

ACC represents 1% of adrenal tumors and has an annual incidence of 0.7–2.0 cases per million. It tends to occur in the first decade (usually before the age of 5 years) or during the fourth and fifth decades of life (median age, 46 years), with a female predilection (1.5:1–2.5:1).

Pheochromocytoma mainly affects adults (mean age, 47 years, with a male predilection) and is found in one per 1000 autopsies (accounting for 3%–7% of incidentaloma).

2.4 Pathogenesis

Risk factors for adrenal tumors include prenatal exposure to carcinogens or fetal alcohol syndrome and inherited tumor disorders (e.g., Li–Fraumeni syndrome [LFS], Beckwith Wiedemann syndrome [BWS], Lynch syndrome, multiple endocrine neoplasia type 1 [MEN1], familial adenomatous polyposis [FAP], neurofibromatosis type 1 [NF1], and Carney complex).

Molecularly, ACC is linked to chromosomal gains and losses; hypermethylation of the promoters of specific genes (e.g., H19, PLAGL1, G0S2, and NDRG2); TP53-inactivating mutations (LFS); insulin-like growth factor (IGF2) overexpression (BWS); mutations in *MENIN* (MEN1), *APC* (FAP), *NF1* (NF1), and *PRKAR1A* (Carney complex) genes; and constitutive activation of the Wnt/β-catenin signaling pathway via activating mutations of the beta-catenin gene (CTNNB1) [3,4].

2.5 Clinical features

Nonfunctional ACA (80%) is largely asymptomatic (no signs of hormonal excess or obvious underlying malignancy). Some ACA (15%) is functional and may produce a pure or mixed endocrine syndrome, ranging from hyperaldosteronism/Conn's syndrome (hypertension, proximal muscle weakness, headache, polyuria, tachycardia with/without palpitation, hypokalemia, hypocalcemia), hypercortisolism/Cushing's syndrome (central obesity, moon facies, plethora, striae, thin skin, easy bruising, hirsutism, telangiectasias, hyperhidrosis) to virilization (females: increased muscle mass, clitoromegaly, facial hair, deep voice, pubic hair; males: penile enlargement, pubic hair; feminization: gynecomastia, impotence).

ACC often manifests as a large abdominal mass, weight loss, fever, cachexia or night sweats, IGF-2–mediated hypoglycemia, hyperreninemic hyperaldosteronism, erythropoietin-associated polycythemia, leukocytosis, and symptoms related to hormone excess (40%–60%) (e.g., glucocorticoid 14%–20%, androgens 5%–20%, oestrogens or mineralocorticoids <5%). However, 20%–30% ACC does not show hormonal overproduction and is discovered incidentally during examination for medical issues unrelated to the adrenal glands (e.g., abdominal pain, fullness, or a palpable mass).

Pheochromocytoma is a hormonally silent incidentaloma associated with nonspecific symptoms (e.g., hypertension, tachycardia, headache, sweating, anxiety, and palpitations).

2.6 Diagnosis

Diagnosis of adrenal tumors involves a medical history review, physical examination (abdominal mass effects, 30%), biochemical tests (glucocorticoid, mineralocorticoid, androgen excess, 40%–60%), imaging studies (CT/MRI abdomen, CT thorax), and pathological examination after tumor removal.

ACA is a unilateral, solitary, golden yellow mass (<5 cm and <50 g) with focal dark areas (due to hemorrhage, lipid depletion, and increased lipofuscin). Functional adenoma may show atrophy in the ipsilateral or contralateral adrenal cortex. ACA appears as a small, homogeneous mass (≤10 HU) on unenhanced CT, with loss of signal on chemical-shift MRI, and a greater contrast washout than adrenal nonadenoma. Microscopically, ACA shows clusters of cells with foamy or enlarged lipid-rich cytoplasm and distinct cell borders; it appears as oncocytic or myxoid.

ACC is an unencapsulated, brown–orange–yellow mass (>5 cm in size, median 11 cm; >100 g in weight) with variegated cut surface (nodularity and fibrous bands due to hemorrhage, cysts, and necrosis), calcification, and invasion of lymphatic channels/blood vessels. ACC appears as a large, heterogeneous enhancing mass of soft tissue attenuation on CT, with an isointense to hypointense signal on T1, a hyperintense signal on T2, a heterogeneous signal drop on chemical shift, and a high 18F-fluorodeoxyglucose (FDG) uptake. Histologically, ACC shows various growth patterns (oncocystic, myxoid, sarcomatoid) of well-differentiated to anaplastic cells (giant cells with hyperchromatic nuclei), capsular invasion, high mitotic activity, and necrosis.

Pheochromocytoma varies from a small, circumscribed to a large, hemorrhagic, and necrotic lesion (mean 7 cm/200 g). A fresh tumor turns dark brown in the presence of potassium dichromate at pH 5–6 (chromaffin reaction). Microscopically, the tumor shows zellballen (small nests or alveolar pattern), trabecular, or solid patterns of polygonal/spindle-shaped cells in a rich vascular network, with finely granular basophilic or amphophilic cytoplasm, oval nuclei, prominent nucleolus, variable inclusion-like structures, and rare mitotic figures.

For differentiation between ACA and ACC, tetrahydro-11-deoxycortisol (THS) excretion >2.35 μmol/24 h is diagnostic of ACC, while a high level of DHEA-S is suggestive of ACA and decreased serum DHEA-S is indictive of adenoma. Using a modified Weiss criteria (5 mitoses per 50 high-powered fields, <25% clear cells, atypical mitotic figures, necrosis, and capsular invasion), ACA (≤5 mitoses per 50 HPF), low-grade ACC (>5 and ≤20 mitoses per 50 HPF), and high-grade ACC (>20 mitoses per 50 HPF) are distinguishable. High-grade ACC is also enriched for mutations of *TP53* or *CTNNB1*, which tend to be mutually exclusive.

The stage of ACC is determined by the tumor–node–metastasis (TNM) classification proposed by the European Network for the Study of Adrenal Tumors (ENSAT), with T1 being a tumor ≤5 cm; T2 being a tumor >5 cm; T3 being tumor infiltration into surrounding (fat) tissue; T4 being tumor invasion into adjacent organs or venous tumor thrombus in vena cava or renal vein; N0 signifying no spread into nearby lymph nodes; N1 signifying positive lymph node(s); M0 signifying no distant metastasis; and M1 signifying the presence of distant metastasis. Thus, Stage I ACC is defined as T1/N0/M0, Stage II ACC as T2/N0/M0, Stage III ACC as T1-2/N1/M0 or T3-4/N0-1/M0, and Stage IV ACC as T1-4/N0-1/M1 [5].

2.7 Treatment

Functional (hormonally active) adrenal tumors are usually treated by surgery, while nonfunctional and non-premalignant ACA (<4 cm, <10 HU) is managed conservatively with long-term follow-up; surgical excision is unnecessary.

For local/locally advanced ACC (limited to adrenal glands), complete surgical resection is the treatment of choice. This may involve open adrenalectomy (tumor >5 cm), laparoscopic adrenalectomy (tumor <5 cm, absence of higher FDG PET uptake), lymph nodes dissection, or a larger surgery involving the adjacent organs [5,6].

For incompletely resectable ACC, debulking surgery is followed by adjuvant mitotane and/or radiotherapy [6,7]. For metastatic ACC (about 50% of cases), treatment options include adrenolytic drug mitotane (alone or in combination with etoposide, doxorubicine, and cisplatin), radiotherapy, and molecular-targeted therapies [epidermal growth factor receptor (EGFR), vascular endothelial growth factor (VEGF), insulin-like growth factor 2 (IGF-2)], which help decrease the risk of recurrence and improve survival marginally [5,6].

Pheochromocytoma is treated by surgery after premedication with adrenergic blockers.

2.8 Prognosis

Prognosis for ACA is good. The percentage of ACC ENSAT stages at diagnosis is 3.3% (Stage I), 37.3% (Stage II), 33.6% (Stage III), and 25.8% (Stage IV). The five-year survival rate for ACC is 66%–82% (Stage I), 58%–64% (Stage II), 24%–50% (Stage III), and 0%–17% (Stage IV) [7]. Post-surgery recurrence is observed in 70%–80% of ACC cases. The most common metastatic sites for ACC are the lungs (40%–80%), liver (40%–90%), and bones (5%–20%) [5].

References

1. Else T, Kim AC, Sabolch A, et al. Adrenocortical carcinoma. *Endocr Rev.* 2014;35(2):282–326.
2. Kaltsas G. Adrenocortical carcinoma. In: De Groot LJ, Chrousos G, Dungan K, et al., editors. *Endotext* [Internet]. South Dartmouth, MA: MDText.com, Inc.; 2000–2015.
3. Lerario AM, Moraitis A, Hammer GD. Genetics and epigenetics of adrenocortical tumors. *Mol Cell Endocrinol.* 2014;386(1–2):67–84.
4. Szyszka P, Grossman AB, Diaz-Cano S, Sworczak K, Dworakowska D. Molecular pathways of human adrenocortical carcinoma—Translating cell signalling knowledge into diagnostic and treatment options. *Endokrynol Pol.* 2016;67(4):427–50.
5. Libé R. Adrenocortical carcinoma (ACC): Diagnosis, prognosis, and treatment. *Front Cell Dev Biol.* 2015;3:45.
6. Creemers SG, Hofland LJ, Korpershoek E, et al. Future directions in the diagnosis and medical treatment of adrenocortical carcinoma. *Endocr Relat Canc.* 2016;23(1):R43–69.
7. Ayala-Ramirez M, Jasim S, Feng L, et al. Adrenocortical carcinoma: clinical outcomes and prognosis of 330 patients at a tertiary care center. *Eur J Endocrinol.* 2013;169(6):891–9.

3 Pancreatic Endocrine Tumors

3.1 Definition

The pancreas is a glandular organ that comprises two functional compartments: the digestive enzyme-secreting exocrine pancreas and the hormone-secreting endocrine pancreas.

Tumors affecting the endocrine pancreas (or pancreatic endocrine tumors [PET]) include glucagonoma, insulinoma, somatostatinoma, gastrinoma, vasoactive intestinal peptide secreting (VIPoma), growth hormone releasing factor secreting (GRFoma), adrenocorticotropin hormone secreting (ACTHoma), PET causing carcinoid syndrome, PET causing hypercalcemia (PTHrPoma), and nonfunctioning PET (PPoma), which together account for 5% of pancreatic tumors (Table 3.1) [1,2].

The most common functional PET types are insulinoma (17%), gastrinoma (15%), VIPoma (2%), glucagonoma (1%), carcinoids (serotonin, 1%), somatostatinoma (1%), PPoma (<1%), ACTHoma (<1%), and GRFoma (<1%); nonfunctional PET accounts for the remainder (30%–40%).

3.2 Biology

Located on the posterior abdominal wall behind the stomach, the pancreas is a J-shaped, flattened, lobulated, soft organ (12–15 cm in length) with five regions (head, uncinate process, neck, body, and tail).

The pancreas consists of two functional compartments: the exocrine and the endocrine. The exocrine pancreas (serous gland) constitutes >95% of the pancreatic parenchyma and includes a million "berry-like" clusters of zymogenic cells (acini) connected by ductules with associated connective tissue, vessels, and nerves. Under the influence of secretin and cholecystokinin, the zymogenic cells secrete trypsin (digesting proteins), lipase (digesting fats), amylase (digesting carbohydrates), and other enzymes, while ductular cells produce bicarbonate, rendering the pancreatic fluid alkaline.

The endocrine pancreas makes up about 2% of pancreatic parenchyma and includes clusters of pancreatic islets (or islets of Langerhans) scattered

Table 3.1 Characteristics of Functioning Pancreatic Endocrine Tumor (PET) Syndromes

Tumor (Syndrome)	Biological Features	Clinical Symptoms	Diagnostic Features	Treatment
Insulinoma (Whipple's triad)	Small solitary pancreatic lesion; rarely malignant (<10%); multiple lesions (MEN); female predominance; median age, fifth decade	Hypoglycemia, neuroglycopenia (confusion, forgetfulness, coma, visual changes, altered consciousness), sympathetic over-stimulation (sweating, tremors, palpitations, weakness, hyperphagia)	Serum glucose <2.5 mmol/L (45 mg/dL); plasma insulin >6 µU/mL (43 pmol/L) by radioimmunoassay (≥3 µU/mL by immunochemoluminescent assay); plasma C-peptide (≥200 pmol/L); sulfonylurea negative; 72-h fast	Resectable tumor (surgery, 95% cure rate); unresectable tumor (diazoxide, octreotide, verapamil, diphenylhydantoin, glucocorticoids, lanreotide, everolimus, mTOR inhibitors); somatostatin analogs
Gastrinoma (Zollinger-Ellison syndrome, or ZES)	Pancreatic/duodenal tumor; MEN 1 association; male predominance; median age, fifth decade; liver metastasis	Severe peptic ulcer (bleeding, obstruction, penetration, perforation), gastroesophageal reflux disease (GERD) (abdominal pain, nausea, heartburn, vomiting), jaundice, secretory diarrhea, hypercalcemia	Fasting serum gastrin (FSG) >1000 pg/mL (normal levels (nl) <100 pg/mL); gastric pH <2; exaggerated response to intravenous secretin ≥120 pg/mL; basal acid output (BAO) >15 mEq/h	Surgical resection; PPIs (omeprazole, lansoprazole, rabeprazole, esomeprazole, pantoprazole); H2 receptor antagonists (cimetidine, ranitidine, nizatidine, famotidine)
VIPoma (Verner–Morrison syndrome, WDHA syndrome, pancreatic cholera)	Solitary pancreatic tumor (>2 cm); female predominance; median age, fifth decade	WDHA (watery diarrhea, hypokalemia, achlorhydria), hyperglycemia, hypercalcemia, flushing	Fasting VIP >200 pg/mL; diarrhea (>3 L per day); liver metastases on somatostatin receptor scintigraphy (SRS or octreoscan)	Fluid/electrolyte replacement; somatostatin analogues (octreotide-LAR, lanreotide-autogel); surgical debulking

(Continued)

Table 3.1 ***(Continued)*** **Characteristics of Functioning Pancreatic Endocrine Tumor (PET) Syndromes**

Tumor (Syndrome)	Biological Features	Clinical Symptoms	Diagnostic Features	Treatment
Glucagonoma	Large pancreatic tumor (>5 cm); metastatic liver lesion	Glucose intolerance, necrolytic migratory erythema/dermatitis, stomatitis/glossitis, hypoaminoacidemia, diabetes, weight loss, diarrhea, anemia	Hyperglucagonemia >500 pg/mL (nl <120); dermatitis; liver metastases and large pancreatic mass	Surgical debulking; somatostatin analogues (octreotide-LAR, lanreotide-autogel); parenteral nutrition; anti-tumor therapies
Somatostatinoma	Pancreatic/duodenal/jejunal tumor	Hyperglycemia, cholelithiasis, steatorrhea, achlorhydria, diabetes, borborygmi, diarrhea, gallstones, weight loss	Somatostatin 15.5 ng/mL (range, 0.16–107 ng/mL) (nl 100 pg/mL); psammoma bodies (psammomatous calcifications)	Nutrition support; hyperalimentation; somatostatin analogues (octreotide-LAR, lanreotide-autogel)
GRFoma	Pancreatic/lung/jejunal/adrenal, foregut/retroperitoneal tumor	Acromegaly	Acromegaly without a pituitary adenoma, with a growth hormone response to TSH or glucose load, or presence of abdominal mass	Somatostatin analogues (octreotide-LAR, lanreotide-autogel); surgical resection; anti-tumor treatments
ACTHoma (Cushing's syndrome)	Pancreatic tumor	Weight gain, abdominal obesity, fatigue, glucose intolerance	Blood cortisol >50 nmol/L (1.81 μg/dL) after dexamethasone administration	Surgery; cortisol antagonists (ketoconazole, metyrapone, mifepristone)

(Continued)

Table 3.1 *(Continued)* Characteristics of Functioning Pancreatic Endocrine Tumor (PET) Syndromes

Tumor (Syndrome)	Biological Features	Clinical Symptoms	Diagnostic Features	Treatment
PET-causing carcinoid syndrome	Pancreatic tumor; liver metastasis; all age groups	Abdominal pain, diarrhea	Positivity for neuron-specific enolase, chromogranin A, and antiserotonin antibodies	Surgical resection; arterial embolization; somatostatin-like substances; serotonin/histamine antagonists; streptozotocin/5-fluorouracil
PET-causing hypercalcemia	Pancreatic tumor; liver metastasis; male predilection	Abdominal pain, anorexia, fatigue, fever, diarrhea	Positivity for chromogranin, synaptophysin, cytokeratins AE1:3, CAM5.2, CK7; negativity for CK20, TTF-1	Cytoreductive intervention, hepatic chemoembolization; somatostatin analogs; octreotide/doxorubicin/capecitabine/temozolomide
Nonfunctional PET (PPoma)	Large pancreatic mass (4–6 cm); liver metastases; fourth or fifth decade	Abdominal pain, jaundice, weight loss, abdominal mass	Positive SRC; plasma pancreatic polypeptide, chromogranin-A, neuron-specific enolase	Surgery/debulking; somatostatin analogues

throughout the pancreas. The islets consist of several cell types (alpha, beta, delta, A, B, C, D, D1, E, F, and enterochromaffin), which are collectively referred to as neuroendocrine cells. It is noteworthy that alpha cells (20% of islets) secrete glucagon; beta cells (68% of islets) secrete insulin and amylin; delta cells (10% of islets) secrete somatostatin; A cells secrete the adrenocorticotropic hormone (ACTH) and melanocyte-stimulating hormone (MSH); D cells secrete gastrin; D1 cells secrete vasoactive intestinal peptide (VIP); E cells (or epsilon cells, <1% of islets) secrete ghrelin; F cells (or PP/gamma cells, 2% of islets) secrete pancreatic polypeptide; and enterochromaffin cells secrete serotonin (5-HT). The secreted hormones are transferred via the blood to control energy metabolism and storage throughout the body.

Previously thought to derive sporadically from the islets of Langerhans (thus the terminology "islet cell tumor"), PET may possibly evolve from pluripotent stem cells (which give rise to islet cells, enterochromaffin cells, argentaffin cells, etc.) in the ductal epithelium of the endocrine pancreas. However, there is evidence suggesting that both mechanisms may give rise to neoplastic lesions. A tumor arising from glucagon-producing alpha cells is known as glucagonoma; a tumor arising from insulin-producing beta cells is known as insulinoma; a tumor arising from somatostatin-producing delta cells is known as somatostatinoma; a tumor arising from gastrin-producing D cells is known as gastrinoma; a tumor arising from pancreatic polypeptide-producing F (PP/gamma) cells is associated with multiple hormonal syndromes; and a carcinoid tumor arises from serotonin-producing cells. In addition, nonfunctioning PET produces no or insufficient hormones but is associated with nonspecific clinical symptoms (e.g., vague abdominal pain) [1,2].

Apart from sporadic occurrence, PET is often associated with inherited genetic disorders, such as gastrinoma (Zollinger–Ellison syndrome) and VIPoma (Verner–Morrison syndrome, pancreatic cholera, watery diarrhea, hypokalemia, and achlorhydria [WDHA] syndrome). Although most PETs are indolent, they have the potential for malignant transformation. Without treatment, PETs may grow and metastasize to the liver.

3.3 Epidemiology

PET accounts for 5% of all pancreatic neoplasms and has an annual incidence of one to five cases per million clinically and a rate of 0.5%–1.5% in autopsy studies. The tumor affects all age groups, with a peak incidence between 30 and 60 years (mean age, 58 years).

3.4 Pathogenesis

PET is associated with inherited genetic syndromes (e.g., multiple endocrine neoplasia type 1 [MEN-1, see Chapter 11], von Hippel–Lindau disease [VHL, see Chapter 12], neurofibromatosis type 1 [NF-1 or Recklinghausen disease VRH], Mahvash disease [see Chapter 9], and tuberous sclerosis complexes 1 and 2 [TSC1/2]) and chromosomal instability (e.g., losses of chromosome 1, p, 6q, 11q, 17p, 21q, and 22q and gains of chromosomes 4, 7, and 9q) [3,4].

3.5 Clinical Features

Early-stage PET with functional cells may cause hormone hypersecretion, leading to specific syndromes (Table 3.1). Late-stage PET with nonfunctional cells may secrete inactive amine and peptide products (e.g., neurotensin, alpha-subunit of human chorionic gonadotropin [alpha-hCG], neuron-specific enolase, pancreatic polypeptide, and chromogranin A], without producing specific syndromes, rather than symptoms related to tumor bulk or metastatic disease (e.g., vague abdominal pain) [3,4].

3.6 Diagnosis

Diagnosis for PET involves endocrine testing, imaging, and histology in order to fully assess the tumor grade, identify the primary and metastatic loci, and determine the functionality of the tumor.

PET usually is a solitary, soft, tan-yellow, well-demarcated mass with a fibrous capsule and necrosis/degeneration (in a larger tumor). Histologically, well-differentiated PET may show a solid nesting pattern, trabecular/gyriform pattern, glandular formation, tubular-acinar, and mixed patterns; stromal fibrosis and amyloid deposition (insulinoma); occasional calcification; psammomatous calcification (somatostatinoma); and cystic, papillary, and angiomatoid/angiomatous patterns as well as PET with ductules. Tumor cells are round to ovoid, with abundant granular eosinophilic cytoplasm, coarsely clumped "salt and pepper" chromatin, discernible nucleoli, and nuclear pseudoinclusions (typical neuroendocrine morphology). Immunohistochemically, PET is positive for synaptophysin, chromogranin, protein gene product 9.5 (PGP9.5), cluster of designation 56, also known as neural cell adhesion molecule (CD56) neuron-specific enolase (NSE), pancreastatin, and neuroendocrine-secretory protein-55 [3,4].

Depending on the level of differentiation, PET may be classified into three categories: (i) well-differentiated grade 1 (<2 mitoses per 10 high power field [HPF]; Ki-67 labeling index <3%), (ii) intermediately differentiated grade 2 (2–20 mitoses per 10 HPF; Ki-67 labeling index 3%–20% and containing small- to medium-sized ovoid nuclei, minimal pleomorphism, and absence of extensive necrosis), and (iii) poorly differentiated grade 3 (also known as pancreatic neuroendocrine carcinoma; >20 mitoses per 10 HPF or Ki-67 index >20% and containing cytological atypia, apparent pleomorphism, and extensive necrosis) [3,4].

The stage of PET is determined according to the TNM system (see Glossary), with stage Ia as T1/N0, stage Ib as T2/N0, stage IIa as T3/N0, stage IIb as T1-3/N1, stage III as T4/any N/M0, and stage IV as any T or N/M1 [3,4].

3.7 Treatment

Surgical resection is the treatment of choice for both functioning and nonfunctioning PET. For metastatic PET, chemotherapy and radiotherapy may be considered for functioning tumors (Table 3.1) [3–5].

Depending on specific circumstances, surgical procedures may range from pancreatectomy, laparoscopic resection, surgical debulking, cytoreductive intervention, radiofrequency ablation (RFA) to microwave ablation.

Chemotherapeutic approaches include systemic therapies (e.g., temozolomide/capecitabine, cisplatin/etoposide, 5-fluorouracil/streptozocin, doxorubicin, chlorozotocin, dacarbazine, sunitinib, everolimus, and interferon alpha), locoregional therapy (e.g., chemoembolization and bland embolization), somatostatin analogs (octreotide and lanreotide), H2 receptor antagonists (cimetidine, ranitidine, nizatidine, and famotidine), proton pump inhibitors (PPIs—omeprazole, lansoprazole, rabeprazole, esomeprazole, and pantoprazole), fluid and electrolyte supplement, and so on. Indeed, systemic therapy is also required for patients with residual disease after surgery and locoregional therapy. Radiotherapy consists of peptide receptor radiotherapy (PRRT) and radioactive polymer microspheres [3–5].

3.8 Prognosis

PET has an average overall five-year survival rate of 55% with localized and resected tumors, but only about 15% with non-resectable tumors. The overall five-year survival rate for nonfunctional PET is 30%.

Important pathological prognostic variables for PET are the presence of tumor necrosis, the mitotic count, and the Ki-67/MIB-1-labeling index.

References

1. Ehehalt F, Saeger HD, Schmidt CM, Grützmann R. Neuroendocrine tumors of the pancreas. *Oncologist*. 2009;14(5):456–67.
2. Asa SL. Pancreatic endocrine tumors. *Mod Pathol*. 2011;24 Suppl 2:S66–77.
3. Ito T, Igarashi H, Jensen RT. Pancreatic neuroendocrine tumors: Clinical features, diagnosis and medical treatment: Advances. *Best Pract Res Clin Gastroenterol*. 2012;26(6):737–53.
4. Ro C, Chai W, Yu VE, Yu R. Pancreatic neuroendocrine tumors: Biology, diagnosis, and treatment. *Chin J Canc*. 2013;32(6):312–24.
5. PDQ Adult Treatment Editorial Board. *Pancreatic Neuroendocrine Tumors (Islet Cell Tumors) Treatment (PDQ®): Health Professional Version*. PDQ Cancer Information Summaries [Internet]. Bethesda, MD: National Cancer Institute (US); 2002–2015.

4
Parathyroid Tumors

4.1 Definition

Tumors affecting the parathyroid glands include parathyroid adenoma (which is the most common form of sporadic parathyroid neoplasia, constituting about 85% of parathyroid malignancies), diffuse hypercellularity (or hyperplasia, which is observed in approximately 15% of cases), and parathyroid carcinoma (or parathyroid cancer, which is responsible for 0.55% of parathyroid malignancies) [1–3].

4.2 Biology

Situated in the neck behind the thyroid gland, the parathyroid glands (or parathyroids) include two pairs of glands, each about the size of a grain of rice (6 mm by 4 mm by 2 mm in dimension, 30 mg in weight). Structurally, the parathyroid glands are composed of chief cells (which are involved in the synthesis and release of parathyroid hormone), oxyphil cells (of unknown function), and adipose cells.

Parathyroid hormone (PTH, also called parathormone) produced by the parathyroids works in synergy with calcitonin secreted by the thyroid gland to keep calcium (and phosphate) homeostasis by releasing calcium from the bone, increasing absorption of calcium from food, and promoting calcium reabsorption by the kidneys.

Hyperparathyroidism results from clonal proliferation of chief cells (a normal response to chronic calcium and vitamin D deficiencies), leading to overproduction of PTH, increased renal resorption of calcium, increased synthesis of 1,24 $(OH)_2D_3$, and increased levels of circulating calcium (hypercalcemia). Featuring an adaptive increase in the synthesis and secretion of PTH, secondary hyperparathyroidism arises as a consequence of hypocalcemia related to chronic renal failure, vitamin D deficiency, and intestinal malabsorption states. Tertiary hyperparathyroidism evolves from preexisting secondary hyperparathyroidism with an autonomous parathyroid hyperfunction.

On the contrary, hypoparathyroidism is caused by damage to the glands or their blood supply during thyroid surgery, leading to underproduction of

PTH and a decreased level of circulating calcium. Pseudohypoparathyroidism results from the resistance of host tissues to the effects of PTH, reducing the blood calcium levels.

Parathyroid adenoma is a benign neoplasm involving one or multiple parathyroid glands, accounting for a majority (85%) of parathyroid tumors and primary hyperparathyroidism.

Parathyroid carcinoma is a rare malignant endocrine cancer, representing 0.55% of parathyroid tumors and 0.4%–5.2% of primary hyperparathyroidism. Parathyroid carcinoma is often associated with severe hypercalcemia and significant mortality. However, <10% of parathyroid carcinoma are nonfunctional and present with normal levels of PTH (so-called normocalcemic hyperparathyroidism).

Diffuse hypercellularity (or hyperplasia) is represented by chief cell hyperplasia and water-clear cell hyperplasia. Showing a diffuse pattern of growth and an absolute increase in parathyroid parenchymal cell mass involving multiple parathyroid glands, primary chief cell hyperplasia often arises sporadically—75% cases are without a known stimulus for PTH hypersecretion, while 25% cases are associated with MEN1, MEN2, familial hypocalciuric hypercalcemia (FHH), neonatal severe hyperparathyroidism, hyperparathyroidism-jaw tumor (HPT-JT) syndrome, familial isolated hyperparathyroidism (FIHPT), and familial hypercalcemic hypercalcuria. However, primary water-clear cell hyperplasia does not have a known association with the heritable hyperparathyroidism syndromes, but it demonstrates a higher degree of hypercalcemia than chief cell hyperplasia [3].

4.3 Epidemiology

Parathyroid tumors have a prevalence of 0.1%–0.4% in the general population and up to 4% in postmenopausal women, and they mainly affect adults (median age, between 45 and 51 years). Although parathyroid adenoma shows a significant female predominance (ratio of 3 to 4:1), parathyroid carcinoma has a roughly equal male to female sex ratio and tends to occur approximately 10 years earlier than parathyroid adenoma.

4.4 Pathogenesis

Risk factors for parathyroid tumors include external radiation exposure (especially parathyroid adenoma), multiple endocrine neoplasia type 1 (MEN1), autosomal dominant familial isolated hyperparathyroidism, HPT-JT syndrome

(15%, which is associated with parathyroid carcinoma or rare multiple, cystic parathyroid adenoma), and a history of thyroid adenoma.

Molecularly, parathyroid tumors are linked to somatic inactivating mutations of the *CDC73/HRPT2* gene (located on chromosome 1q31, encoding parafibromin, a tumor suppressor involved in HPT-JT syndrome and 70% of sporadic parathyroid carcinoma), *MEN1* gene (on chromosome 11q13, encoding menin, which is involved in MEN1 syndrome and up to 40% of sporadic parathyroid adenoma), and the *PRUNE2* gene (on chromosome 9q21.2, encoding Ras homolog suppressor; losses of chromosomal DNA [1p21-p22, 4q24, 6q22-q24, 9p21-pter, 11q, 11p, 13q14-q31, 15q and 17] and gains of chromosomal DNA [1q31-q32, 5, 7, 9q33-qter, 16p, 19p, and Xc-q13]); mutations in the *CDKN1B*/p27 cyclin-dependent kinase inhibitor (CDKI) gene (noted in 5% of sporadic adenoma but overexpression of cyclin D1 in up to 40% of adenoma); hypermethylation of the promoter region of genes (e.g., *CDKN2B*/p15, *CDKN2A*/p16, *SFRPs, RASSF1, HIC1,* and *APC* as well as genes encoding molecules of the Wnt/β-catenin pathway); aberrant expression of chromatin regulatory molecules (e.g., upregulation of histone H1.2, H2A, and H2B genes in the histone microcluster on chromosome 6p22.2-p21.3); and altered microRNA expression profiles (e.g., downregulation of miR-296-5p, miR-139-3p, miR-126-5p, miR-26b, and miR-30b, and upregulation of miR-222, miR-503, and miR-517c) [4,5].

4.5 Clinical features

Clinically, nonfunctioning parathyroid malignancies (including 10% parathyroid carcinoma) often present with an expanding mass on the neck, hoarseness, and dysphagia, whereas functional parathyroid malignancies may show additional symptoms related to hypercalcemia, including dehydration, hypotension, mental confusion, muscle weakness, fatigue, joint and bone pain, nausea, anorexia, weight loss, polyuria, polydipsia, kidney stones, constipation, and dyspepsia (typical of primary hyperparathyroidism).

In particular, parathyroid carcinoma (cancer) is characterized by severe hypercalcemia (albumin-corrected serum calcium >3 mmol/L, often >14 mg/dL, commonly between three and ten times the upper limit of the normal range), a palpable neck mass (>3 cm), and a third/second generation PTH assay ratio (>1), unilateral vocal cord paralysis, hyperparathyroid bone disease, and renal involvement (e.g., nephrolithiasis or nephrocalcinosis). Parathyroid carcinoma tends to be localized in the inferior parathyroid glands and rarely metastasizes to other sites (e.g., regional lymph nodes <5% and distant sites <2%).

4.6 Diagnosis

Clinicopathologically, parathyroid tumors are defined by (i) vascular invasion, (ii) perineural invasion, (iii) gross invasion into adjacent anatomical structures, and (iv) metastasis.

Parathyroid adenoma is a soft, round-oval, tan, or reddish-brown mass approximately 1.5 cm in size and <0.5 g in weight (with tumors weighing <0.1 g referred to as microadenomas) and located predominantly in the superior parathyroid glands. Foci of cystic change and formation of a fibrous capsule may occur in large adenoma (often in association with HPT-JT syndrome). Microscopically, parathyroid adenoma contains chief cells (pure chief cell variant in 50% of cases; other cases with transitional oncocytes or vacuolated chief cells) arranged in cords, nests, sheets, and follicles and frequently forming a palisaded structure around blood vessels. The tumors cells have large, round, densely stained, hyperchromatic, and pleomorphic nuclei; large nucleoli (endocrine atypia); sometimes prominent follicular architecture; and mitotic activity (in up to 70% of cases; Ki-67 value of <4%). Other parathyroid adenoma variants include oncocytic adenoma (with abundant granular eosinophilic cytoplasm, 11% of cases), lipoadenoma (hamartoma, with branching cord surrounded by mature fat cells with areas of fibrosis and lymphocytic infiltration), and clear cell adenoma (with small nuclei, dense chromatin, and abundant vacuolated cytoplasm). Immunohistochemically, the tumor cells are positive for cytokeratins, PTH, and chromogranin A, but they are negative for thyroglobulin and thyroid transcription factor-1 [6].

Parathyroid carcinoma is a firm, grayish white, stony-hard, lobulated, poorly circumscribed, and relatively large mass between 3.0 and 3.5 cm in size, with ill-defined borders and a dense, fibrous, grayish-white capsule that infiltrates adjacent tissues. On ultrasound, parathyroid carcinoma often shows a hypoechoic lump with irregular margins in the front of the neck. Biochemically, it has high serum calcium (>14 mg/dL) and PTH (especially if twice the normal value). Microscopically, parathyroid carcinoma shows sheets or lobules of tumor cells separated by dense fibrous bands, mitotic figures (5 per 50 high-power fields; Ki-67 values of 6%–8.4%), focal necrosis, capsular or vascular invasion, in addition to stromal calcifications, round to oval hyperchromatic nuclei, granular chromatin, inconspicuous nucleoli, and scanty eosinophilic cytoplasm. Immunohistochemical staining for parafibromin helps confirm parathyroid carcinoma. Loss of nuclear and/or nucleolar expression of parafibromin (coded by the CDC73/HRPT2 gene) represents a diagnostic, prognostic, and predictive biomarker for parathyroid carcinoma [7,8].

4.7 Treatment

Treatment of choice for parathyroid tumors is surgery (e.g., parathyroidectomy alone, en bloc resection of the cancer with surrounding tissue, and oncological resection, including the field of lymphatic drainage), along with preoperative medical management to lower elevated calcium levels and to correct other metabolic disturbances due to hyperparathyroidism. For parathyroid carcinoma, en bloc dissection of the tumor with the thyroid lobe, the ipsilateral parathyroid, and any other affected tissue leads to the best prognosis as local excision (parathyroidectomy) alone is associated with increased recurrence (51% vs. en bloc resection 8%). Accessible recurrent or metastatic metastases should be resected when possible, and debulking of functional carcinomas may help reduce PTH production [9].

Adjuvant radiation with 40–50 Gy is recommended in patients at high risk for local relapse (e.g., evidence of microscopic residual disease after surgery or positive surgical margins).

Hypercalcemia-related symptoms in functional tumors may be controlled by cinacalcet HCl (which activates the calcium-sensing receptor, CASR, on parathyroid cells and inhibits PTH release), bone antiresorptive agents (e.g., bisphosphonates, denosumab, plicamycin, calcitonin, gallium pamidronate, and mithramycin), and calcimimetic agents (which directly block secretion of PTH from the parathyroid glands). This is particularly relevant for patients with unresectable disease or without measurable disease [9].

Techniques useful for localization of parathyroid tumors include technetium Tc 99m sestamibi scan, single photon emission computed tomography, computed tomography-(99m)Tc-sestamibi (CT-MIBI) image fusion, ultrasound, CT, MRI, selective angiogram, and selective venous sampling for PTH.

4.8 Prognosis

Parathyroid carcinoma is generally a slow-growing tumor with low malignant potential and has a 5- and 10-year overall survival of 85% and 49%–77%, respectively.

Nonfunctional parathyroid carcinoma has a poorer prognosis due to the delayed detection at advanced stages, relative inefficiency of adjuvant treatment modalities, and frequent relapse and distant metastasis (in the lungs, bone, and lymph nodes).

References

1. DeLellis RA. Parathyroid tumors and related disorders. *Mod Pathol.* 2011;24 Suppl 2:S78–93.
2. Hu J, Ngiam KY, Parameswaran R. Mediastinal parathyroid adenomas and their surgical implications. *Ann R Coll Surg Engl.* 2015;97(4):259–61.
3. Wang L, Han D, Chen W, et al. Non-functional parathyroid carcinoma: A case report and review of the literature. *Canc Biol Ther.* 2015;16(11):1569–76.
4. Costa-Guda J, Arnold A. Genetic and epigenetic changes in sporadic endocrine tumors: Parathyroid tumors. *Mol Cell Endocrinol.* 2014;386(1–2):46–54.
5. Verdelli C, Corbetta S. Epigenetic alterations in parathyroid cancers. *Int J Mol Sci.* 2017;18(2):pii:E310.
6. Neagoe RM, Sala DT, Borda A, Mogoantă CA, Műhlfay G. Clinicopathologic and therapeutic aspects of giant parathyroid adenomas—Three case reports and short review of the literature. *Rom J Morphol Embryol.* 2014;55(2 Suppl):669–74.
7. Dytz MG, Souza RG, Lázaro AP, et al. Parathyroid carcinoma and oxyphil parathyroid adenoma: An uncommon case of misinterpretation in clinical practice. *Endocr J.* 2013;60(4):423–9.
8. Duan K, Mete Ö. Parathyroid carcinoma: Diagnosis and clinical implications. *Turk Patoloji Derg.* 2015;31 Suppl 1:80–97.
9. PDQ Adult Treatment Editorial Board. *Parathyroid Cancer Treatment (PDQ®): Health Professional Version.* PDQ Cancer Information Summaries. Bethesda, MD: National Cancer Institute (US); 2002–2017.

5
Pituitary Tumors

Jian Yi Li

5.1 Definition

Among tumors of the pituitary gland, pituitary adenomas (which are benign, non-spreading neuroendocrine tumors that are separated into pituitary adenoma and atypical pituitary adenoma) are most common, while pituitary carcinomas (which are cancerous and spreading) may occasionally affect older people. Other rare tumors/lesions occurring in the pituitary region include craniopharyngioma, Rathke cleft cyst, germ cell tumors (germinoma, teratoma, and choriocarcinoma), Langerhans cell histiocytosis, and gangliocytoma (Table 5.1).

5.2 Biology

Located just below the brain and above the nasal passages, the pituitary gland is connected directly to a part of the brain called the hypothalamus and thus provides a link between the brain and the endocrine system. Considered the master control gland, the pituitary makes hormones that control the levels of hormones made by most other endocrine glands in the body.

The pituitary gland is divided into two parts: (i) the larger, front part is known as the anterior pituitary (which makes hormones that regulate growth, thyroid gland, adrenal gland, testis, and ovaries) and (ii) the smaller, back part is known as the posterior pituitary (which, as an extension of brain tissue from the hypothalamus, makes hormones that regulate the kidneys, uterus, and breasts).

Pituitary tumors usually arise from secretory cells in the pituitary adenohypophysis (the anterior pituitary) and rarely develop in the posterior pituitary.

5.3 Epidemiology

Pituitary tumors, accounting for 10%–25% of all intracranial tumors, mainly occur in the fourth and seventh decades, with female preponderance [1,2]. During postmortem examination, the rate of incidental pituitary tumors is

Table 5.1 Tumors of Sellar and Suprasellar Regions

Tumors of Sellar and Suprasellar Regions	
Neuroendocrine tumors	Pituitary adenoma Atypical pituitary adenoma Pituitary carcinoma
Non-neuroendocrine tumors TTF-1 expressing tumors	Gangliocytoma Pituicytoma Granular cell tumor Spindle cell oncocytoma
Tumors of non-pituitary origin	Craniopharyngioma • Adamantinomatous (96% activating CTNNB1 mutation, β-catenin nuclear stain) • Papillary (95% with BRAF V600F) Langerhans cell histiocytosis (>50% with BRAF V600F mutation) Meningioma Chordoma Germ cell tumor Metastatic tumors

approximately 10% [1]. Prolactin (PRL)-secreting and adrenocorticotropic hormone (ACTH)-secreting adenomas are the most common pituitary adenomas.

5.4 Pathogenesis

Although obvious recurrent somatic gene mutation in pituitary adenomas is not identified, genetic alterations in the following genes may play a role in the tumorigenesis of pituitary adenomas.

- *Activation of oncogenes*: Alpha subunit of the guanine nucleotide stimulatory protein (Gs-alpha) gene, Ras, the pituitary tumor transforming gene-1 (PTTG-1), and fibroblast growth factor receptor-4(FGFR-4) [3–6]
- *Inactivation of anti-oncogenes*: Retinoblastoma (Rb), p16, p27, and p53 [1]

Hormonal factors also play an important role in the development of pituitary adenoma [1].

- *Overproduction of stimulatory hormones*: Growth hormone-releasing hormone (GHRH), corticotropin-releasing hormone (CRH), thyrotrophin-releasing hormone (TRH), or gonadotropin-releasing hormone (GnRH) [1]
- *Defect in inhibitory hormone regulation*: Dopamine, somatostatin (SS), glucocorticoid hormones, thyroid hormones, or gonadal hormones [1]

Pituitary adenoma can be seen in a few inherited familial syndromes such as multiple endocrine neoplasia type 1 (MEN1), Carney complex, familial isolated pituitary adenoma (FIPA), and McCune-Albright syndrome [7].

5.5 Clinical features

Clinical features from functional pituitary adenomas (often microadenomas, <10 mm) are due to the overproduction of hormones.

Prolactin-secreting adenomas (prolactinomas) can cause amenorrhea, galactorrhea, and infertility in women but cause decreased libido, impotence, and rarely galactorrhea in men.

Growth hormone (GH)-secreting adenomas can cause gigantism in children and adolescents but cause acromegaly in adults.

ACTH-secreting pituitary adenomas can cause Cushing's disease, which is manifested by weight gain, centripetal obesity, moon face, violet striae, and easy bruisability.

Thyroid-stimulating hormone (TSH)-secreting adenomas can cause hyperthyroidism.

Clinical features from nonfunctional pituitary adenomas (usually macroadenomas, >1.0 mm) are due to the mass effect and compression on surrounding structures.

They include visual defects, nonspecific headaches, pituitary hormone deficiencies, diplopia, ptosis, ophthalmoplegia, cerebrospinal fluid (CSF) rhinorrhea, hypothalamic dysfunction, and obstructive hydrocephalus. Pituitary apoplexy manifested as excruciating headaches and diplopia is caused by sudden hemorrhage into the adenoma.

5.6 Diagnosis

5.6.1 Hormone tests

Serum prolactin level is elevated in prolactinomas. Serum insulin-like growth factor 1 (IGF-1) level is a better test for acromegaly caused by GH-secreting adenoma than serum growth hormone (GH) level. ACTH-secreting adenoma can cause increased 24-h urine cortisol levels. TSH-secreting adenomas can cause elevated serum TSH, T3, and T4 levels. Follicle-stimulating hormone (FSH) or luteinizing hormone (LH) levels can be increased in FSH or LH-secreting adenomas, respectively.

5.6.2 Imaging studies

CT scan is less sensitive and specific than MRI, but it is much better for detecting calcification than MRI.

Microadenoma (<10 mm): On T1-weighted MRI images, pituitary microadenomas usually exhibit slightly isointense to normal glands. On T2-weighted images, they often are slight hyperintense to normal glands. On T1 post-contrast images, dynamic sequences show a rounded region of delayed enhancement compared to normal glands, while delayed images demonstrate hypo-enhancement to isointense to normal glands to hyperintense (retained contrast) [8].

Macroadenoma (>10 mm): On T1-weighted MRI images, macroadenomas show isointense to gray matter, are often heterogeneous, and vary in signal due to areas of cystic change/necrosis/hemorrhage. On T2-weighted images, macroadenomas are isointense to gray matter and often heterogeneous, and they vary in signal due to areas of cystic change/necrosis/hemorrhage. On T1 post-contrast images, solid components exhibit moderate to bright enhancement. Hemorrhage can be better seen on a T2 gradient echo [9]. MRI also can assess the relationship between the tumor and adjacent structures, such as the optic chiasm, cavernous sinus, brain parenchyma, cranial nerve, and blood vessels.

5.6.3 Histological findings

A normal pituitary gland consists of a mixture of different cell types, which form small acini surrounded by reticulin meshwork (Figure 5.1). Pituitary adenoma is composed of monomorphic cells (usually one cell type) with small, round nuclei, salt-and-pepper chromatin, and abundant cytoplasm without reticulin meshwork among tumors cells (Figure 5.2). The growth pattern includes diffuse, trabecular, pseudo-acinar, or pseudo-papillary with abundant vascularity in the tumor. Mitosis is rare. Tumor cells are positive for neuroendocrine markers such as synaptophysin and chromogranin and low molecular keratin (CAM5.2). Immunohistochemical stains for GH, PRL, ACTH, LH, FSH, and TSH are available for further characterizing pituitary adenomas. In order to make the diagnosis of atypical adenoma, the Ki-67 labeling index and p53 positivity rate should be ≥3%.

Pituitary carcinomas can, but do not always, show obvious nuclear pleomorphism with prominent nucleoli, high mitotic activity, or tumor necrosis. Pituitary carcinomas are uncommon and comprise less than 1% of pituitary tumors. A diagnosis of pituitary carcinoma can only be made when there is central nervous system (CNS) dissemination or distant metastasis.

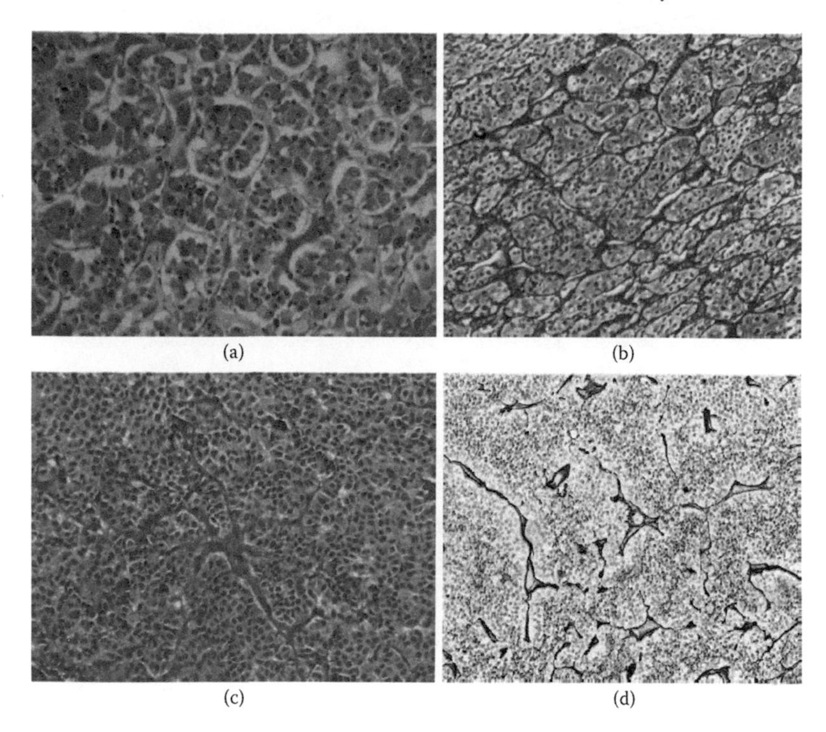

Figure 5.1 The comparison between normal pituitary gland and pituitary adenoma. (a) Normal pituitary gland is composed of a mixture of different cell types, 100X, hematoxylin and eosin (H&E) stain. (b) Reticulin stain on normal pituitary gland shows small acini surrounded by reticulin meshwork. 40X (c) Pituitary adenoma consists of sheets of monomorphic tumor cells, 100X, H&E stain. (d) Reticulin stain on pituitary adenoma demonstrates the complete disruption of reticulin meshwork, 40X.

5.6.4 Differential diagnoses

Differential diagnoses for pituitary adenomas and carcinomas include rare tumors/lesions that also occur at the suprasellar and sellar regions (Table 5.1).

Adamantinomatous craniopharyngioma is a cystic mass lesion that commonly affects the pediatric population and exhibits adamantinomatous epithelium at the periphery, loose "stellate reticulum" in the center, and "wet keratin" with or without calcification. Papillary cranioparyngioma is mainly found in adults and resembles squamous papilloma.

Pituicytoma is a rare benign tumor derived from pituitcytes (modified glial cells) in neurohypophysis or pituitary infundibulum, possesses bland spindle cells in dense fascicles or storiform growth pattern, and stains positive for

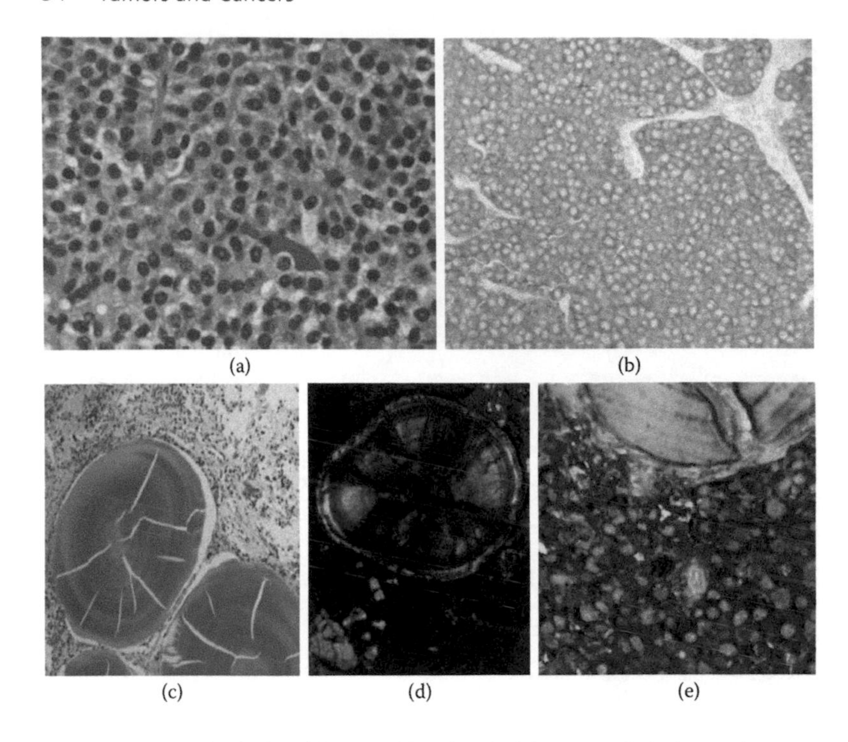

Figure 5.2 Histological and immunohistochemical features of pituitary adenoma. (a) Pituitary adenoma contains tumor cells with abundant granular cytoplasm and perivascular arrangement. The tumor cell nuclei are relatively monotonous with "salt-and-pepper" chromatin, 400X, hematoxylin and eosin (H&E) stain. (b) Immuohistochemical stain shows that tumor cells are positive for synatophysin, 200X. (c) Prolactinoma with hyaline spheroids, 40X, H&E stain. (d) Hyaline spheroid in prolactinoma exhibits apple-green under polarized light on Congo red-stained section, 40X. (e) Immuohistochemical stain shows that prolactioma tumor cells are positive for prolactin, 400X.

vimentin, S100 protein, and thyroid transcription factor-1 (TTF-1); but, it is negative for neuroendocrine markers (i.e., synaptophysin) and pituitary hormones.

A granular cell tumor is thought to arise from pituicytes from the pituitary infundibulum and posterior lobe. It contains polygonal to elongate cells with granular, eosinophilic cytoplasm (lysosome rich) and small, eccentric nuclei and stains positive for S100 protein and TTF-1 as well as variably positive for CD68; but, it is negative for synaptophysin and pituitary hormones.

A spindle cell oncocytoma is possibly derived from folliculostellate cells in the pituitary anterior lobe, has spindle to epithelioid tumor cells with eosinophilic granular mitochondria-rich cytoplasm, and reacts with vimentin, epithelial membrane antigen (EMA), S100 protein, anti-mitochondrial antibodies, and TTF-1; but, it is negative for synaptophysin and pituitary hormones.

5.7 Treatment

5.7.1 Surgery

Endoscopic trans-sphenoidal minimally invasive surgery is the treatment of choice for almost all pituitary tumors.

5.7.2 Medicine

Prolactin-secreting adenomas can be treated with dopamine agonists (e.g., bromocriptine and cabergoline).

GH- and TSH-secreting adenomas can be treated with somatostatin analogs (Octreotid). Growth hormone receptor antagonists (Somavert[R], pegvisomant) can be used for GH-secreting adenomas.

Temozolomide can effectively treat aggressive tumors, including pituitary carcinoma [10].

5.7.3 Radiation

Radiotherapy can be used for patients with incompletely resected or recurrent tumors or for patients who have severe medical conditions and cannot tolerate surgery.

Patients with pituitary carcinomas should be treated with local resection of the primary tumor and metastatic foci followed by radiation therapy. These patients require close postoperative follow-up.

5.8 Prognosis

Overall, the prognosis for both pituitary adenoma and atypical pituitary adenoma is very good, especially after gross total resection. However, they can recur even after gross total resection. Long-term clinical follow-up is recommended.

The prognosis for pituitary carcinoma is usually poor. The mean survival is 2 years and ranges from 1/4 year to 8 years. However, a few long-term survivors have been documented.

References

1. Asa SL, Ezzat S. The cytogenesis and pathogenesis of pituitary adenomas. *Endocr Rev.* 1998;19:798–827.
2. Ezzat S, Asa SL, Couldwell WT, et al. The prevalence of pituitary adenomas: A systematic review. *Cancer.* 2004;101:613–9.
3. Ezzat S, Zheng L, Zhu XF, Wu GE, Asa SL. Targeted expression of a human pituitary tumor-derived isoform of FGF receptor-4 recapitulates pituitary tumorigenesis. *J Clin Invest.* 2002;109:69–78.
4. Landis CA, Masters SB, Spada A, Pace AM, Bourne HR, Vallar L. GTPase inhibiting mutations activate the alpha chain of Gs and stimulate adenylyl cyclase in human pituitary tumours. *Nature.* 1989;340:692–6.
5. Vallar L, Spada A, Giannattasio G. Altered Gs and adenylate cyclase activity in human GH-secreting pituitary adenomas. *Nature.* 1987;330:566–8.
6. Vlotides G, Eigler T, Melmed S. Pituitary tumor-transforming gene: Physiology and implications for tumorigenesis. *Endocr Rev.* 2007;28:165–86.
7. Vasilev V, Daly A, Naves L, Zacharieva S, Beckers A. Clinical and genetic aspects of familial isolated pituitary adenomas. *Clinics.* (*Sao Paulo*) 2012;67 Suppl 1:37–41.
8. Bickle IA, Gaillard F. *Pituitary microadenoma*, Radiopaedia.org. 2016. http://radiopaedia.org/articles/pituitary-microadenoma. Accessed September 22, 2016.
9. Knipe H, Weerakkody Y. *Pituitary macroadenoma*, Radiopaedia.org. 2016. http://radiopaedia.org/articles/pituitary-macroadenoma-1. Accessed September 22, 2016.
10. Syro LV, Ortiz LD, Scheithauer BW, et al. Treatment of pituitary neoplasms with temozolomide: A review. *Cancer.* 2011;117:454–62.

6
Thyroid Tumors

6.1 Definition

Tumors affecting the thyroid gland include (i) *epithelial tumors* (thyroid adenomas and related tumors [follicular adenomas, hyalinizing trabecular tumors]; thyroid carcinomas [papillary carcinomas, follicular carcinomas, poorly differentiated carcinomas, anaplastic/undifferentiated carcinomas, squamous cell carcinomas, mucoepidermoid carcinomas, sclerosing mucoepidermoid carcinomas with eosinophilia, mucinous carcinomas, medullary thyroid carcinomas, mixed medullary and follicular cell carcinomas, spindle cell tumors with thymus-like differentiation, carcinomas showing thymus-like differentiation]); (ii) *nonepithelial tumors* (teratomas, primary lymphomas and plasmacytomas, ectopic thymomas, angiosarcomas, smooth muscle tumors, peripheral nerve sheath tumors, paragangliomas, solitary fibrous tumors, Rosai-Dorfman disease, follicular dendritic cell tumors, Langerhans cell histiocytosis); and (iii) *secondary tumors*.

The most common primary thyroid tumors are papillary thyroid carcinoma (PTC), including classical (PTC-CV) and follicular (PTC-FV) variants and representing 75%–85% of all thyroid malignancies; follicular thyroid carcinoma (FTC), representing 10%–20%; medullary thyroid carcinoma (MTC), representing 5%; and anaplastic (undifferentiated) thyroid carcinoma (ATC), representing 1%–2% of all thyroid malignancies [1,2].

6.2 Biology

Situated at the base of the throat anterior to the trachea, the thyroid is a brownish-red organ composed of two elongated, wing-shaped lobes (5–6 cm in length, 25–30 g in weight) connected by a median isthmus. A conical pyramidal lobe often ascends from the adjacent part of either lobe (or the isthmus) toward the hyoid bone.

Covered by an outer visceral fascia and an inner capsule, the thyroid gland comprises lobes and lobules that are divided by septae. The lobules include follicles (the structural units of the gland) made up of a layer of epithelium enclosing a colloid-filled cavity.

The thyroid epithelium consists of two cell types: follicular cells (also called thyroid epithelial cells or thyrocytes) and parafollicular cells (also called C, clear, or light cells). Originating from a median endodermal mass in the region of the tongue (foramen cecum), follicular cells are primary cells lining the follicles and are responsible for production and secretion of the colloid (an iodinated glycoprotein or iodothyroglobulin—a precursor of thyroid hormones, which ultimately matures to play a key role in control of the heart rate, blood pressure, body temperature, and basal metabolic rate). Scattered along the basement membrane of the thyroid epithelium, parafollicular cells arise from the fourth pharyngeal pouch and are involved in the generation of the hormone calcitonin (a protein central to calcium homeostasis).

Although PTC, FTC, and ATC all originate from thyroid follicular cells, PTC and FTC are considered well-differentiated tumors with a relatively favorable prognosis, while ATC is a poorly differentiated tumor that probably evolves as an anaplastic transformation of a papillary, follicular, or Hürthle cell carcinoma. A papillary microcarcinoma (or occult papillary tumor) is a subset of PTC that measures ≤1 cm [2].

MTC develops from thyroid parafollicular (C) cells of the ultimobranchial body of the neural crest, demonstrates neuroendocrine features, and has a poorer prognosis than PTC and FTC. Although 70%–80% of MTCs are sporadic, 20%–30% are familial and associated with multiple endocrine neoplasia 2A (MEN2A), multiple endocrine neoplasia 2B (MEN2B), and familial medullary thyroid carcinoma (FMTC). Familial MTC tends to affect younger patients (peak age, 35 years) than does sporadic MTC (peak age, 40–60 years).

6.3 Epidemiology

Thyroid carcinoma (cancer) is the most common malignancy of the endocrine system, accounting for 1%–3% of all tumors worldwide. It has an annual incidence of 9–12 cases per 100,000 and shows a bias for women (female to male ratio of 3:1). Although all age groups are affected, adults between 45 and 54 years (mean age, 50 years at diagnosis) are particularly vulnerable. The prevalence is higher in the white and Asian/Pacific Islander populations than in other populations.

6.4 Pathogenesis

Risk factors for thyroid carcinoma include past exposure to ionizing radiation in the head and neck region (particularly for PTC), family history of thyroid cancer or thyroid disease, familiar genetic disorders (e.g., multiple endocrine neoplasia type 2, particularly for MTC), and genetic mutations (e.g., BRAF-V600E, *RAS, PTEN, CTNNB1, TP53, IDH1, NDUFA13* [*GRIM19*], *RET,* and PAX8/PPARG) [3].

6.5 Clinical features

Thyroid carcinoma often presents with a nodule in the thyroid region of the neck, an enlarged lymph node, pain, difficulty swallowing, difficulty breathing, hoarseness or a change in voice, and adenopathy. MTC is associated with excess calcitonin.

6.6 Diagnosis

Diagnosis of thyroid tumors involves medical history review (past head and neck radiation, thyroid disease, and cancer), physical examination, laboratory investigation (serum thyroid-stimulating hormone [TSH] to differentiate between functional and nonfunctional nodules and also between indolency and malignancy, anti-TPO/anti-Tg antibody, serum calcitonin to exclude MTC), imaging studies (thyroid ultrasonography, radionuclide scanning, CT, MRI, and PET scan to determine hyperfunctioning [hot], isofunctioning [warm], or nonfunctioning [cold] nodule), and cytologic or histologic examination (fine-needle aspiration, large-needle biopsy, core-needle biopsy) [2,4].

Papillary thyroid carcinoma (PTC) is a solid, firm, gray-white, often multifocal (20%), encapsulated (10%), or infiltrative, well-differentiated tumor with central scar, cysts, fibrosis, and calcification. Being typically smaller and less aggressive than follicular thyroid carcinoma (FTC), PTC may be found in the thyroid only (67%), in the thyroid and cervical nodes (13%), and in the nodes only (20%). Microscopically, PTC demonstrates complex, branching papillae with fibrovascular cores and dense fibrosis. Cuboidal cells lining papillae contain nuclei (larger than follicular nuclei) that are overlapping with finely dispersed optically clear chromatin (so-called ground glass, Orphan Annie nuclei), micronucleoli, eosinophilic intranuclear inclusions, and nuclear longitudinal grooves. Psammoma bodies (due to tumor cell necrosis) are present in the papillary stalk in fibrous stroma (50% of cases). Several histologic variants (e.g., follicular, tall cell, diffuse sclerosis, solid, and oncocytic) are identifiable. Immunohistochemically, PTC is positive for CK19 (strong and diffuse), HBME1 (recommended to differentiate from benign mimics), CK7, high molecular weight keratin, *RET,* thyroglobulin (less intense than follicular neoplasms), TTF1, S100, EMA, vimentin, estrogen receptors, c-kit (weak), alpha-1-antitrypsin (90%), p63 (82%), thyroid peroxidase (50%), galectin3, carcinoembryonic antigen (CEA) (occasional); but, it is negative for CK20. Molecularly, PTC may harbor BRAF point mutations (30%), RET rearrangements at 10q11.2 (45%), and RAS point mutations (follicular variant) [5].

Follicular thyroid carcinoma (FTC) is a gray-tan-pink, solitary, encapsulated, well-differentiated tumor with focal hemorrhage, variable fibrosis, and calcification. Microscopically, FTC shows follicular differentiation with follicular cell invasion of tumor capsule and/or blood vessels, but it lacks papillary nuclear characteristics and psammoma bodies. FTC is associated with iodine deficiency and has a tendency to undergo hematogenous metastases to lungs and bone—but rarely to regional lymph nodes. Histologic variants of FTC include Hürthle cell (oncocytic), clear cell, and insular (poorly differentiated) carcinomas. Hürthle cell carcinoma (oxyphilic cell carcinoma) displays oncocytic characteristics (polygonal shape, eosinophilic granular cytoplasm, hyperchromatic or vesicular nuclei, large nucleoli, and abundant mitochondria) in >75% of follicular cells and may have a poorer prognosis than usual FTC. Immunohistochemically, FTC is positive for thyroglobulin, low molecular weight keratin, EMA, and TTF1. Molecularly, FTC contains Ras mutations (49%) and PAX8-PPAR gamma rearrangements (36%).

Medullary thyroid carcinoma (MTC) is a gray-tan-yellow, firm, solid, typically nonencapsulated, poorly differentiated tumor without a well-formed capsule. Although sporadic MTC is usually solitary, most of the familial MTCs have bilateral, multicentric foci. Microscopically, MTC shows round, spindle, or polyhedral C cells in nests, cords, or follicles defined by sharply outlined fibrous bands (which separate tumors into nodules); tumor cells with uniform round or oval nuclei, punctate (salt-and-pepper) nuclear chromatin, and granular cytoplasm; amyloid deposits from calcitonin in stroma; presence of mucin (42%); and C cell hyperplasia (in familial but not sporadic MTC). Immunohistochemically, MTCs are positive for calcitonin, Congo red (for amyloid), CEA, low molecular weight keratin, chromogranin A and B, synaptophysin, neuron-specific enolase, TTF1, and progesterone receptors; but, it is negative for thyroglobulin and estrogen receptors. Molecularly, MTCs contain activating mutations of RET (different from those in PTC).

Anaplastic (undifferentiated) thyroid carcinoma (ATC) (also called carcinosarcoma or sarcomatoid carcinoma) is a rapidly enlarging, bulky, solid, extremely invasive, poorly differentiated neck mass with necrosis and hemorrhage. Due to its early metastasis to the surrounding lymph nodes and distant sites, ATC is a fatal disease (almost 100% disease-specific mortality) that is responsible for 40% of thyroid cancer deaths despite representing only <2% of thyroid cancer cases. Microscopically, ATC shows three patterns: (i) large, pleomorphic giant cells resembling osteoclasts with cellular connective tissue septa and cavernous blood-filled sinuses resembling aneurysmal bone cysts; (ii) spindle cells resembling sarcoma; and (iii) squamoid cells (relatively undifferentiated or epithelial with occasional focal keratinization). Vascular invasion and mitotic figures are common, while rhabdoid inclusions are rare. Immunohistochemically, an ATC is positive for keratin (not in the sarcomatoid variant); vimentin (spindle cell component); p53 (well-differentiated tumor, usually p53 negative); CD68 (osteoclast-like cells), Ki67, PCNA, and PAX8 (79%); EMA and CEA (variable); but, it is negative for thyroglobulin, TTF1, bcl2, calcitonin, desmin, and muscle specific actin. Molecularly, ATC harbors BRAF mutations (50%) and N-RAS (31%) [6].

The stage of thyroid carcinoma is determined according to the TNM classification system (see Glossary) designed by the American Joint Committee on Cancer (AJCC). PTC and FTC are usually classified as stages I, II, III, and IV; MTC as stages 0, I, II, III, IVA, IVB, and IVC; and virtually all ATC is diagnosed as stage IV [4].

6.7 Treatment

For localized, stage I, II, and some III thyroid carcinoma, the first-line treatment is surgery (e.g., total thyroidectomy for high-risk tumors ≥1 cm and lobectomy for tumors <1 cm), which may be followed by adjunct radiotherapy (e.g., ^{131}I ablation, especially for primary papillary or follicular tumors of 1–4 cm with intermediate to high risk of recurrence, and external beam radiation therapy, especially for unresectable, recurrent tumors and if ^{131}I uptake is minimal) and/or chemotherapy (e.g., sorafenib, sunitinib, vandetanib, and cabozantinib, in addition to doxorubicin plus cisplatin for ATC). For stage IV or metastatic thyroid carcinoma, partial resection (of ^{131}I-nonesponsive metastases), radiotherapy, and/or palliative chemotherapy are recommended [4,7]. However, ATC appears to be resistant to most treatments and is usually fatal (mean survival, 6 months).

Thyroid carcinoma follow-up involves yearly chest X-rays and check of thyroglobulin levels (especially for a well-differentiated carcinoma after total thyroidectomy). A post-thyroidectomy high serum thyroglobulin increase (>10 ng/mL) with TSH stimulation level is indicative of recurrence.

6.8 Prognosis

Thyroid carcinoma has an overall 5-year survival rate of 85% for females and 74% for males. For PTC, the 5-year survival rates for stage I, II, III, and IV are

100%, 100%, 93%, and 51%, respectively; for FTC, 100%, 100%, 71%, and 50%, respectively; for MTC, 100%, 98%, 81%, and 28%, respectively; for ATC, 7% (stage IV). The 10-year survival rates for PTC, FTC, and MTC are 93%, 85%, and 75%, respectively [8].

References

1. Eloy C, Ferreira L, Salgado C, Soares P, Sobrinho-Simões M. Poorly differentiated and undifferentiated thyroid carcinomas. *Turk Patoloji Derg*. 2015;31 Suppl 1:48–59.
2. Dideban S, Abdollahi A, Meysamie A, Sedghi S, Shahriari M. Thyroid papillary microcarcinoma: Etiology, clinical manifestations, diagnosis, follow-up, histopathology and prognosis. *Iran J Pathol*. 2016;11(1):1–19.
3. Papp S, Asa SL. When thyroid carcinoma goes bad: A morphological and molecular analysis. *Head Neck Pathol*. 2015;9(1):16–23.
4. Nguyen QT, Lee EJ, Huang MG, Park YI, Khullar A, Plodkowski RA. Diagnosis and treatment of patients with thyroid cancer. *Am Health Drug Benefits*. 2015;8(1):30–40.
5. PathologyOutlines.com website. *Papillary carcinoma—General*. http://www.pathologyoutlines.com/topic/thyroidpapillary.html. Accessed May 12, 2017.
6. PathologyOutlines.com website. *Anaplastic carcinoma*. http://www.pathologyoutlines.com/topic/thyroidUndiff.html. Accessed May 12, 2017.
7. Grewal RK, Ho A, Schöder H. Novel approaches to thyroid cancer treatment and response assessment. *Semin Nucl Med*. 2016;46(2):109–18.
8. Glikson E, Alon E, Bedrin L, Talmi YP. Prognostic factors in differentiated thyroid cancer revisited. *Isr Med Assoc J*. 2017;19(2):114–8.

7 Carney Complex

7.1 Definition

First described in 1985 by J. Aidan Carney as a complex of myxomas, spotty skin pigmentation (lentigines), and endocrine overactivity, Carney complex (CNC) is a rare autosomal dominant syndrome linked to mutations in the *PRKAR1A* gene and with a heightened risk of developing various endocrine and non-endocrine tumors and lesions.

The most notable endocrine tumors and lesions associated with CNC include primary pigmented nodular adrenocortical disease (PPNAD), adrenocorticotropic hormone (ACTH)-independent Cushing's syndrome, growth hormone (GH)- or growth hormone and prolactin (GH-PRL)-secreting (mammosomatroph) pituitary adenomas, thyroid nodules, large-cell calcifying Sertoli cell tumors (LCCSCTs), adrenal rest tumors (ectopic PPNAD), Leydig cell tumors in males, and ovarian cystadenomas in females. On the contrary, common non-endocrine tumors/lesions are presented by myxomas (heart, breast, skin); lentigines (hamartomatous melanocytic lesions); blue nevi; compound nevi; café-au-lait spots; psammomatous melanotic schwannomas (PMSs); adrenal, hepatocellular, and pancreatic cancer; and bone tumors (osteochondromyxomas) [1,2].

7.2 Biology

Chromosomes are thread-like molecules that carry hereditary information from parents to offspring. Humans possess 22 nonsex chromosome pairs (also called autosomes) and two sex chromosomes (two X chromosomes in females but an X chromosome and a Y chromosome in males).

Produced by meiosis cell division, human gametes (egg and sperm) contain half the number of chromosomes as diploid parent cells; that is, each diploid parent cell has 23 chromosome pairs (46 chromosomes in total), while an egg or a sperm has a single set of 23 chromosomes (23 chromosomes in total). During fertilization, the sperm combines with the egg to form a zygote, which again contains two sets of 23 chromosomes.

When a dominant gene located on one of the 22 nonsex chromosomes (autosomes) from one parent is mutated, there is a 50% chance that an affected child will inherit the mutated gene (dominant gene) from this parent and a normal gene (recessive gene) from the other parent, whereas an unaffected child will have two normal genes (recessive genes) and will not develop or pass on the disease. Autosomal dominant inheritance refers to the transfer of one copy of the altered gene in each cell that is sufficient to cause the disorder.

CNC (also known as Carney syndrome) is considered an autosomal dominant disorder involving germline mutations or large deletions in the gene coding protein kinase cAMP-dependent type I regulatory subunit alpha (PKAR1A) located on 17q22-24. However, about 20% of CNC cases are sporadic without known affected relatives, and these appear to arise through *de novo* germline mutations.

CNC should not be confused with Carney triad, which is a distinct disease entity consisting of a triad of gastrointestinal stromal tumors, pulmonary chondroma, and extra-adrenal paraganglioma.

7.3 Epidemiology

CNC is a rare disorder, with about 750 affected individuals identified to date. There is a female bias (60% females and 40% males) among CNC patients. In addition, transmission of CNC through a female affected parent is almost fivefold more frequent than through a male due possibly to the fact that male patients often harbor LCCSCTs that may cause infertility. The median age of diagnosis is 20 years, although the penetrance of CNC is 70%–80% by the age of 40 years and >95% by the age of 50 years [1,2].

PPNAD demonstrates a bimodal age distribution: a first peak during infancy and a second peak between the second and third decade of life, with a median age of 34 years at diagnosis and a female predilection (sex ratio 2.4:1).

Cardiac myxoma occurs in any chamber and has a median age of 20 years at diagnosis, in contrast to sporadic cardiac myxoma, which is found exclusively in the left atrium and frequently affects older females.

7.4 Pathogenesis

The majority (80%) of CNC cases occur in a familial context resulting from inactivating germline mutations or large deletions in the type 1α regulatory subunit of protein kinase A (*PRKAR1A*) gene situated at the 24.2–24.3

locus of the long arm of chromosome 17 (i.e., 17q22–24). The *PRKAR1A* gene encompasses 11 exons and encodes type 1 alpha subunit—a protein of 384 amino acids that helps turn on or off protein kinase A (which promotes cell growth and division). Mutations in the *PRKAR1A* gene lead to an abnormal type 1 alpha subunit that is quickly degraded by the cell. In the absence of type 1 alpha, protein kinase A remains turned on more often than normal, contributing to uncontrolled cell proliferation. Pathogenic mutations (>125 identified to date) in the *PRKAR1A* gene range from single base substitutions (e.g., c.82C>T, c.491_492delTG, and c.709-2_709-7 delATTTTT), small (≤15 bp) deletions/insertions, and combined rearrangements spreading along the whole open reading frame (ORF) of the gene, to large deletions covering most of the exons or the whole gene locus. These alterations contribute to frame shifts and/or premature stop codons (PSC), resulting in shorter or otherwise defective *PRKAR1A* mRNAs, which are degraded by the nonsense-mediated mRNA decay (NMD) surveillance mechanism instead of being encoded to protein. Additional components of the complex associated with defects of other PKA subunits (e.g., the catalytic subunits PRKACA on chromosome 19p13.1 [adrenal hyperplasia] and PRKACB on chromosome 1p31.1 [pigmented spots, myxoma, pituitary adenoma]) have been identified. Further, a 10 Mb region in the 2p16 locus has been also implicated in CNC pathogenesis [3–5].

Some 20% of CNC cases do not harbor mutations in the *PRKAR1A* gene but have genetic abnormality in a specific region on the short (p) arm of chromosome 2 (i.e., 2p16) [3–5].

7.5 Clinical features

The most common symptoms of CNC are skin lesions (up to 70% of CNC cases), cardiac myxoma (20%–40%), and PPNAD (25% to 60%).

Abnormal skin pigmentation (lentigines, blue nevus, café-au-lait spots) occurring at birth usually represent the first manifestation of CNC. Subsequent clinical presentations include PPNAD leading to Cushing's syndrome and cardiac myxoma (cardiomyopathy, cardiac arrhythmia, emboli/stroke, sudden death), then LCCSCT and thyroid nodules (cystic or nodular disease, benign thyroid adenoma, thyroid cancer), followed by acromegaly, pituitary tumors (somatomammotroph hyperplasia, GH-producing adenoma with acromegaly, prolactinoma), pancreatic tumors (acinar cell carcinoma, adenocarcinoma, intraductal pancreatic mucinous neoplasia), ovarian tumors (ovarian cyst, serous cystadenoma, cystic teratoma), breast tumors (breast and nipple myxoma, myxoid fibroadenoma, ductal adenoma), PMS, and osteochondromyxomas [1,2].

7.6 Diagnosis

Diagnosis of CNC on the basis of 12 major clinical criteria and two supplemental criteria has a sensitivity of 98%. Molecular testing of the *PRKAR1A* gene provides confirmation in about 60% of cases.

The 12 major clinical criteria include (i) spotty skin pigmentation with a typical distribution (the lips, conjunctiva and inner or outer canthi, vaginal and penile mucosa); (ii) blue nevus, epithelioid blue nevus (multiple); (iii) cutaneous and mucosal myxomas; (iv) cardiac myxomas; (v) breast myxomatosis or fat-suppressed MRI findings; (vi) osteochondromyxoma; (vii) PPNAD or a paradoxical positive response of urinary glucocorticosteroids to dexamethasone administration during Liddle's test; (viii) acromegaly due to GH-producing adenoma or evidence of excess GH production; (ix) LCCSCT or characteristic calcification on testicular ultrasonography; (x) thyroid carcinoma or multiple, hypoechoic nodules on thyroid ultrasonography—in a young patient; (xi) psammomatous melanotic schwannoma; and (xii) breast ductal adenoma [1–2].

The two supplemental criteria are (i) affected first-degree relative and (ii) inactivating mutation of the *PRKAR1A* gene or activating pathogenic variants of *PRKACA* (single base substitutions and copy number variation) and *PRKACB* [1,2].

A diagnosis of CNC is made when a patient has either two of the major clinical criteria confirmed by histology, imaging, or biochemical testing, or one major clinical criterion and one supplemental criterion. Therefore, a negative test for mutations in the *PRKR1A* gene does not exclude CNC in an individual who meets clinical criteria. If a *PRKR1A* gene mutation is detected, genetic screening is recommended for first-degree relatives (parents, siblings, and offspring) [1,2].

Differential diagnoses for CNC lentigines include benign familial lentiginosis, Peutz–Jeghers syndrome, LEOPARD syndrome, Noonan syndrome with lentiginosis, and Bannayan–Riley–Ruvalcaba syndrome (*PTEN* hamartoma tumor syndrome); those for CNC café au lait spots are McCune–Albright syndrome, neurofibromatosis type 1 (NF1), neurofibromatosis type 2 (NF2), and Watson syndrome; that for CNC cardiac myxoma is sporadic myxoma cardiomyopathy; those for CNC thyroid tumors are Cowden syndrome (*PTEN* hamartoma tumor syndrome) and sporadic thyroid tumors; that for LCCSCTs is Peutz-Jeghers syndrome; those for adrenal cortical tumors are Beckwith-Wiedemann syndrome (BWS), Li–Fraumeni syndrome (LFS), multiple endocrine neoplasia type 1 (MEN1), congenital adrenal hyperplasia resulting from 21-hydroxylase deficiency, and McCune–Albright syndrome (MAS); those for GH-secreting pituitary adenomas (somatotropinomas) are MEN1 and isolated familial somatotropinomas (IFS); and those for CNC schwannoma are NF1, NF2, and isolated familial schwannomatosis [1–2].

7.7 Treatment

As CNC is associated with a spectrum of clinical diseases, therapeutic approaches vary greatly from one specific complication/tumor to another.

Cardiac, cutaneous, and mammary myxomas require surgical resection; PPNAD may be treated by bilateral adrenalectomy or with inhibitors of steroidogenesis (e.g., ketoconazole or mitotane) in selected cases; boys with LCCSCT may develop gynecomastia, premature epiphyseal fusion, and induction of central precocious puberty, and they should be treated with orchiectomy and/or aromatase inhibitors; a Leydig cell tumor is treated by inguinal orchiectomy; SH and/or GH-producing pituitary adenomas may be treated by transsphenoidal/transcranial surgery or with somatostatin analogues; thyroid nodules may be treated surgically; and operable PMS is treated by surgery to remove primary and/or metastatic lesions [1,2].

Patients with CNC should be followed closely for clinical manifestations at least once a year, involving (i) annual echocardiogram and biannual cardiac imaging for patients with a history of cardiac myxoma; (ii) regular skin evaluations; (iii) blood tests for serum GH, prolactin, and IGF-1, and urine tests for urinary free cortisol (UFC), for acromegaly in adolescent patients; (iv) thyroid gland (neck) clinical examinations; (v) adrenal computed tomography for PPNAD; pituitary MRI and MRI of brain, spine, chest, abdomen, retroperitoneum, and pelvis for PMS; (vi) testicular ultrasonography in males for LCCSCT; (vii) transabdominal ultrasound of the ovaries for females; and (viii) close monitoring of linear growth rate and annual pubertal staging in pre-pubertal children [1,2].

7.8 Prognosis

Patients with CNC have an average life span of 50–55 years. Common causes of death are related to complications of heart myxoma (e.g., emboli/strokes, post-operative cardiomyopathy, and cardiac arrhythmias), metastatic PMS, and pancreatic and other tumors.

Genetic counseling should emphasize that if a parent of the index case is affected, the sibling has a 50% risk of CNC. However, if the index case harbors a *de novo* mutation, the sibling has a 1% risk of CNC. Further, the child of an affected individual has a 50% chance of acquiring CNC. Prenatal testing for CNC is possible by chorionic villous sampling (CVS) at 10–12 weeks of gestation or amnioparacentesis at 15–18 weeks of gestation [1,2].

References

1. Kaltsas G, Kanakis G, Chrousos. Carney's complex. In: De Groot LJ, Chrousos G, Dungan K, et al, editors. *Endotext* [Internet]. South Dartmouth, MA: MDText.com, Inc.; 2000–2013.
2. Correa R, Salpea P, Stratakis CA. Carney complex: An update. *Eur J Endocrinol.* 2015;173(4):M85–97.
3. Salpea P, Stratakis CA. Carney complex and McCune Albright syndrome: An overview of clinical manifestations and human molecular genetics. *Mol Cell Endocrinol.* 2014;386(1–2):85–91.
4. Berthon AS, Szarek E, Stratakis CA. PRKACA: The catalytic subunit of protein kinase A and adrenocortical tumors. *Front Cell Dev Biol.* 2015;3:26.
5. Schernthaner-Reiter MH, Trivellin G, Stratakis CA. MEN1, MEN4, and Carney complex: Pathology and molecular genetics. *Neuroendocrinology.* 2016;103(1):18–31.

8 Li-Fraumeni Syndrome

8.1 Definition

First described by Frederick Li and Joseph Fraumeni in 1969 in a group of families with an unusually high incidence of cancers, Li–Fraumeni syndrome (LFS) is a rare autosomally dominant syndrome associated with germ line TP53 mutations and early development of soft tissue sarcoma, osteosarcoma, premenopausal breast cancer, brain tumor, adrenocortical carcinoma, leukemia, and lung cancer as well as other neoplasms [1,2].

There exists a similar condition called Li–Fraumeni-like syndrome (LFL), which also displays an increased risk for developing multiple cancers in childhood but has a different pattern of specific cancers in comparison with classic LFS [1,2].

8.2 Biology

Compared to other hereditary cancer syndromes that are often linked to one (or a few) specific type(s) of cancer, LFS is a clinically heterogeneous hereditary cancer predisposition syndrome characterized by the clustering pattern of a spectrum of core tumors (i.e., bone and soft-tissue sarcomas, central nervous system tumors, leukemia, adrenocortical carcinoma, and breast cancer) in children and adolescents.

LFS appears to arise in the background of germline TP53 gene mutations (70%–80% of cases). It is possible that genetic damage in the tumor suppressor gene TP53 induces loss of heterozygosity (LOH) and/or gain of function (GOF), facilitating malignant transformation (the so-called two-hit model) [3,4].

However, around 20%–30% of patients with LFS do not carry TP53 mutations. Instead, non-TP53 mutations (involving the checkpoint kinase 2 [CHEK2] gene on 22q12.1 and a single-nucleotide polymorphism 309 in the MDM2 gene, which encodes a key negative regulator of TP53) have been found [2].

8.3 Epidemiology

LFS is a rare, autosomal, dominant, hereditary cancer syndrome that affects about one in 5,000 individuals in Europe and North America, with about 500 LFS families recorded in the database of the International Association for Research on Cancer (IARC).

LFS-related cancers show a bimodal distribution, with one peak before the age of 10 and a second peak between the ages of 30 and 50 years. The most common tumors during childhood and early adulthood are osteosarcoma, adrenocortical carcinoma, glioma, and soft tissue sarcoma as well as other malignancies; whereas breast cancer (a 90% risk in women by the age of 60 years) and soft tissue sarcomas are frequently identified in adults. Due to the increased risk of breast cancer in females, cancer penetrance is 93% for female carriers compared with 73% for male carriers. Among the TP53 mutation carriers, the risk for developing cancer by ages 20, 30, 40, and 50 years is estimated to be 12%, 35%, 52%, and 80%, respectively [2].

Within LFS-affected families, the prevalence of specific neoplasms differs during childhood, adolescence, and adulthood, with soft tissue sarcomas, brain tumors, and ACC commonly observed in zero to 10 years of age, bone sarcomas in 11–20 years of age, and breast cancer and brain tumors in >20 years of age. Males without childhood cancers may develop multiple primary tumors in their 50s. Patients with LFL are generally older than those with LFS.

It appears that there is a reverse relationship between the age at diagnosis of the first primary tumor and the relative risk of developing a second primary tumor among patients with LFS and LFL, with the relative risk of developing a second primary tumor estimated at 83%, 9.7%, 1.5%, and 5.3% for patients aged 0–19, 20–44, ≥45, and all ages, respectively, at diagnosis of first primary tumor.

8.4 Pathogenesis

LFS is an autosomal dominant disorder for which one copy of the altered *TP53* gene in each cell is sufficient to pass the disease to 50% of offspring. Not surprisingly, an affected person often has a parent and other family members with a similar cancer profile. As a tumor suppressor gene located on the short arm of chromosome 17 (i.e., 17p13.1), *TP53* spans 22 kb, including 11 exons, and encodes a 53 kDa tumor protein (TP53) that exerts an integrated antiproliferative impact through control of the cell cycle, apoptosis, senescence, DNA repair, differentiation, and basal energy metabolism. Mutations in *TP53* thus allow cells to divide in an uncontrolled manner.

To date, more than 700 germline mutations have been reported in the *TP53* gene, with most mutations scattered throughout the gene, including hotspots at codons 125, 158, 175, 196, 213, 220, 245, 248, 273, 282, and 337. These mutations can be categorized into two groups: (i) gain-of-function mutations (missense mutations) that show a dominant-negative effect or promote an oncogenic effect and that appear to be implicated in an earlier onset of cancer and (ii) loss-of-function mutations such as nonsense and frameshift mutations [5–7].

Interestingly, codon 337 (p.Arg337His) is frequently detected in ACC and CPC among Brazilian patients. Further, missense mutations at codons 164–194 and codons 237–250 are found in brain cancer, while those at codons 115–135, S2-S2-H2 motif: codons 273–286 are observed in ACC. In addition, the missense mutation p.Val31Ile is noted in patients with a late onset of cancer, and frameshift mutation p.Pro98Leufs*25 is detected in patients with an early onset of cancer.

Mutations in another tumor suppressor gene, *CHEK2* on 22q12.1, also have been identified in some families with LFS, although carriers of *CHEK2* mutations show lower incidence of cancer in their families than carriers of TP53 mutations. In addition, the association between *MDM2* SNP309 and increased cancer risk appears to be modest, and the interaction between *MDM2* SNP309 and *TP53* mutation is not statistically significant.

8.5 Clinical features

Patients with LFS often present with cancer-related symptoms such as loss of appetite; headaches; pains; new moles, lumps, or swellings; changes in vision or nerve function; and unexplained weight loss.

Specific cancers related to LFS include sarcomas (e.g., rhabdomyosarcoma, osteosarcoma, leiomyosarcomas, liposarcomas, and histiosarcoma, representing 25%–30% of all LFS-associated tumors), brain tumors (e.g., astrocytoma, glioblastoma, medulloblastoma, ependymoma, choroid plexus carcinoma, and malignant triton tumor, representing 9%–16%), adrenocortical carcinoma (ACC, representing 10%–14%), breast cancer (representing 25%–30%), and other cancers (e.g., colorectal, esophageal, pancreatic, and stomach cancers; renal cell carcinoma; endometrial, ovarian, prostate, and gonadal germ cell tumors; leukemia, Hodgkin and non-Hodgkin lymphomas; lung cancer; melanoma and non-melanoma skin cancers; and non-medullary thyroid cancer) [1,2].

8.6 Diagnosis

Diagnosis of LFS is based on the established clinical criteria or detection of a germline pathogenic variant in *TP53* irrespective of family cancer history [1].

Clinical criteria for classic LFS include (i) a sarcoma diagnosed before the age of 45 years, (ii) a first-degree relative (parent, sibling, or child) with any cancer before the age of 45 years, and (iii) a first- or second-degree relative (grandparent, aunt/uncle, niece/nephew, or grandchild) with any cancer before the age of 45 years or a sarcoma at any age. Individuals meeting all three criteria are considered as having LFS. For those who only meet some of these criteria, a positive *TP53* test is sufficient because *TP53* is the only gene definitively associated with LFS [1,2,8].

Clinical criteria for LFL are used for affected families who do not meet classic criteria for LFS. These consist of two suggested definitions.

LFL definition 1 stipulates (i) a person diagnosed with any childhood cancer, sarcoma, brain tumor, or adrenal cortical tumor before age 45; (ii) a first-degree or second-degree relative diagnosed with a typical LFS cancer (e.g., sarcoma, breast cancer, brain cancer, adrenal cortical tumor, or leukemia) at any age; and/or (iii) a first-degree or second-degree relative diagnosed with any cancer before age 60.

LFL definition 2 stipulates two first-degree or second-degree relatives diagnosed with a typical LFS cancer (e.g., sarcoma, breast cancer, brain cancer, adrenal cortical tumor, or leukemia) at any age [1].

Because 80% of families with features of LFS have an identifiable *TP53* pathogenic variant, molecular detection of *TP53* mutations provides a valuable tool for confirmation of LFS cases. By sequencing the entire *TP53* coding region (exons 2–11), about 95% of *TP53* pathogenic variants (including many missense variants) can be identified.

Differential diagnoses for LFS include hereditary breast–ovarian cancer syndrome (which tends to harbor a pathogenic variant in *BRCA1* or *BRCA2* instead of *TP53* as germline *TP53* pathogenic variants account for <1% of total breast cancer cases) and constitutional mismatch repair deficiency syndrome (CMMR-D syndrome, which is associated with childhood leukemia, brain tumors, or early-onset gastrointestinal cancer, but is caused by the inheritance of two mutated alleles of a mismatch repair gene, including *MLH1*, *MSH2*, *MSH6*, *PMS1*, and *PMS2*) [1].

8.7 Treatment

The management of LFS centers on treatment of manifestations, prevention of primary manifestations and secondary complications, and surveillance.

Prophylactic mastectomy rather than lumpectomy is recommended for *TP53* positive women with breast cancer. This will help reduce the risk of a second primary breast tumor and avoid radiation therapy. Colonoscopy may be considered for surveillance as well as primary prevention of colorectal cancer.

Drugs currently under development for treating LFS-associated tumors include monoclonal antibodies (e.g., trastuzumab [herceptin] for breast and stomach cancer), small molecules (e.g., PhiKan083, PRIMA-1, CP31398, WR1065, MIRA-1, STIMA-1, RETRA, Nutlin -3, and RITA), and metformin (a drug that activates the *TP53*-AMPK pathway and preferentially inhibits growth of cells lacking functional *TP53*).

Surveillance measures for children or adults with a germline *TP53* pathogenic variant consist of (i) comprehensive annual physical examination, (ii) prompt evaluation of lingering symptoms and illnesses (e.g., headaches, bone pain, or abdominal discomfort) by a physician, (iii) annual breast MRI and twice annual clinical breast examination from age 20 to 25 years, and (iv) routine screening for colorectal cancer with colonoscopy every 2–3 years from age 25 years.

Individuals with germline *TP53* pathogenic variants should avoid known carcinogens (e.g., sun exposure, tobacco use, occupational exposures, and excessive alcohol intake) and minimize exposure to diagnostic and therapeutic radiation [1].

8.8 Prognosis

Patients with LFS have a 50% risk of developing cancer before the age of 30 and a 90% lifetime risk for developing any type cancer.

Genetic counseling for LFS-affected individuals should focus on the fact that LFS is an autosomal dominant, hereditary cancer predisposition syndrome with a germline *TP53* pathogenic variant in 80% of cases. Offspring of an affected individual will have a 50% chance of inheriting the deleterious allelic variant. On the contrary, about 7% of patients with LFS/LFL-spectrum tumors diagnosed at a young age have *de novo* mutation, and their siblings will be at a very low risk for the condition.

References

1. Schneider K, Zelley K, Nichols KE, Garber J. Li-Fraumeni Syndrome. In: Pagon RA, Adam MP, Ardinger HH, et al, editors. *GeneReviews®* [Internet]. Seattle, WA: University of Washington, Seattle; 1993–2017. 1999 [updated 2013 Apr 11].
2. Malkin D. Li-Fraumeni syndrome. *Genes Canc.* 2011;2(4): 475–84.
3. Sorrell AD, Espenschied CR, Culver JO, Weitzel JN. Tumor protein p53 (TP53) testing and Li-Fraumeni syndrome: Current status of clinical applications and future directions. *Mol Diagn Ther.* 2013;17(1):31–47.
4. Pantziarka P. Primed for cancer: Li Fraumeni syndrome and the pre-cancerous niche. *Ecancermedicalscience.* 2015;9:541.
5. Ariffin H, Hainaut P, Puzio-Kuter A, et al. Whole-genome sequencing analysis of phenotypic heterogeneity and anticipation in Li-Fraumeni cancer predisposition syndrome. *Proc Natl Acad Sci U S A.* 2014;111(43):15497–501.
6. Giacomazzi J, Selistre SG, Rossi C, et al. Li-Fraumeni and Li-Fraumeni-like syndrome among children diagnosed with pediatric cancer in Southern Brazil. *Cancer.* 2013;119(24):4341–9.
7. Park KJ, Choi HJ, Suh SP, Ki CS, Kim JW. Germline TP53 mutation and clinical characteristics of Korean patients with Li-Fraumeni syndrome. *Ann Lab Med.* 2016;36(5):463–8.
8. Giacomazzi CR, Giacomazzi J, Netto CB, et al. Pediatric cancer and Li-Fraumeni/Li-Fraumeni-like syndromes: A review for the pediatrician. *Rev Assoc Med Bras.* 2015;61(3):282–9.

9 Mahvash Disease

9.1 Definition

Mahvash disease is a newly discovered, rare autosomal recessive hereditary disorder caused by homozygous inactivating mutations in the glucagon receptor gene, leading to elevated glucagon levels (hyperglucagonemia without symptoms of glucagonoma syndrome), pancreatic α-cell hyperplasia, pancreatic neuroendocrine tumors (PNETs), and mild hypoglycemia [1].

9.2 Biology

Glucagon (abbreviated from its original name: *glu*cose *agon*ist) is a 29-amino acid polypeptide of 3.485 kDa generated from the cleavage of proglucagon by proprotein convertase 2 in pancreatic islet α cells, which are the main endocrine cell populations (33%–46%) that coexist in the islet of Langerhans along with insulin-secreting β-cells (48%–59%), in addition to scarce δ- and poly-peptide releasing (PP)-cells (<10%). Alternate products of proglucagon generated in intestinal L cells are glicentin, GLP-1 (an incretin), IP-2, and GLP-2 (promotes intestinal growth) [2].

The coordinated secretion of glucagon and insulin by α- and β-cells of the islet of Langerhans, respectively, is vital for the regulation of glycaemia. While hypoglycemic conditions induce α-cell secretion of glucagon, hyperglycemic conditions increase β-cell release of insulin.

As a counter-regulatory hormone for insulin, which lowers the extracellular glucose (which is usually stored in the liver as a polymer of glucose molecules, or polysaccharide glycogen), glucagon works to raise the concentration of glucose in the bloodstream through its binding to the glucagon receptor (GCGR, a G protein-coupled receptor of 485 amino acids) located in the plasma membranes of hepatocytes (liver cells) as well as in the kidney, pancreas, heart, brain, and smooth muscle. This activates the stimulatory G protein and ten adenylate cyclase, triggering cAMP production and converting stored polysaccharide glycogen into glucose (so-called glycogenolysis). After the exhaustion of stored glycogen, glucagon promotes synthesis of additional glucose (called gluconeogenesis) in the liver and kidneys. In the meantime, glucagon also shuts off glycolysis in the liver, turning glycolytic intermediates into gluconeogenesis [2].

Factors that are known to stimulate glucagon secretion include hypoglycemia, epinephrine, arginine, alanine, acetylcholine, and cholecystokinin, whereas factors that inhibit glucagon secretion are somatostatin, insulin, PPARγ/retinoid X receptor heterodimer, increased free fatty acids and keto acids in the blood, and increased urea production.

Pancreatic tumors, such as glucagonoma (an amino acid deficiency syndrome), are known to induce abnormally elevated levels of glucagon, with symptoms of necrolytic migratory erythema, reduced amino acids, and hyperglycemia.

Mahvash disease (named after the first name of the index patient and also known as nesidioblastosis, alpha cell hyperplasia, microglucagonoma, or nonfunctioning islet cell tumor) was initially noted in 2007 in a 60-year-old female patient who presented with abdominal pain, mild hypoglycemia (735 pg/mL vs. normal, <311 pg/mL) and elevated glucagon levels (59,284 pg/mL vs. normal, <150 pg/mL) without symptoms of glucagonoma syndrome (e.g., skin rash, hyperglycemia, weight loss, thromboembolism, or stomatitis). Normal levels of insulin, gastrin, vasoactive intestinal peptide, and calcitonin were observed. This patient had a past history of gastroesophageal reflux, duodenal ulcer, diverticulosis, anemia, β thalassemia minor, anxiety, depression, and meningioma. An abdominal CT showed a hypertrophic pancreas and a 3-cm mass in the uncinate process. Histological examination of the surgically resected pancreas revealed a nonfunctioning neuroendocrine tumor composed of an apparently normal pancreas and numerous glucagon-positive, insulin-negative hyperplastic islets in the walls of the pancreatic ducts, along with two microadenomas, suggesting α cell neogenesis and high lifetime recurrence risk. Sequencing analysis of the 13 coding exons and intronexon borders of the glucagon receptor gene (*GCGR*) identified a homozygous, missense C-to-T mutation in exon 4, resulting in replacement of proline with serine at amino acid residue 86 (P86S) in the GCGR protein. Although the index patient is homozygous for P86S mutation, her brother has two wild-type *GCGR* alleles, suggesting that their parents are heterozygous for P86S mutation [3,4].

By 2015, more than 10 cases of Mahvash disease from several families had been identified in both the United States and Germany, most of which have confirmed biallelic mutations in the glucagon receptor and have demonstrated an autosomal recessive inheritance pattern. In contrast to an autosomal dominant disorder such as Li-Fraumeni syndrome, in which a mutated gene is inherited from one parent (see Chapter 8), an autosomal recessive disorder involves inheritance of two mutated genes, one from each parent. In this scenario, two carriers each with one mutated gene (recessive gene) and one normal gene (dominant gene) have a 25% chance of getting an unaffected child with two normal genes, a 50% chance of getting an unaffected carrier child, and a 25% chance of getting an affected child with two recessive genes. It is likely that a lack of negative feedback from glucagon receptor signaling results in compensatory hyperplasia of pancreatic α-cells and secondary tumorigenesis in Mahvash disease [1].

In a Gcgr–/– mouse model of Mahvash disease, the pancreas is apparently normal in gross appearance and islet morphology at birth, and it develops α cell hyperplasia without obvious dysplasia by two to three months. This is followed by dysplastic islet growth by five to seven months and microscopic PNET by 12 months. Occasionally, well-differentiated PNET in Gcgr–/– mice may metastasize to the liver. Besides significantly reduced survival, Gcgr–/– mice demonstrate decreased fetal survival and vision loss. Remarkably, the pancreas pathology of Gcgr–/– mice is identical to that of humans with Mahvash disease, and Gcgr–/– mice provide further evidence of the autosomal recessive mode of inheritance of Mahvash disease [1].

9.3 Epidemiology

Mahvash disease is a rare autosomal recessive, familial, PNET. Based on the known frequency (0.1%) of two potentially inactivating exonal single nucleotide polymorphisms (T76M and G442V), it is estimated that the frequency of inactive *GCGR* mutation may occur at approximately 0.2%. Thus, the prevalence of Mahvash disease is about four cases per million and that of the heterozygous carrier state approximately 0.4%.

From more than 10 cases of Mahvash disease reported to date, affected patients develop gross PNET in their middle age, with severe hyperglucagonemia before and after tumor resection. The patients do not have parathyroid or pituitary tumors or a family history of multiple endocrine neoplasia syndrome type 1 (MEN1). Recurrence is not evident two to nine years postoperatively [1].

9.4 Pathogenesis

The inactivating glucagon receptor mutations increase the number of the pancreatic alpha cells and facilitate subsequent transition from hyperplasia to PNET, which is the only neoplasm known to occur in patients with Mahvash disease.

The progression from hyperplasia to PNET appears to be a multistep pathogenesis. It is possible that homozygous inactivating mutation of the *GCGR* gene incapacitates the glucagon receptor, sending signals to the pancreatic alpha cells to proliferate (i.e., pancreatic α cell hyperplasia, which is largely a hormonally-driven α cell neogenesis) and produce more glucagon. However, as the mutated glucagon receptor has greatly reduced activity compared with the normal glucagon receptor, a patient with Mahvash disease has lowered glucose levels and does not develop glucagonoma syndrome. PNET tumorigenesis in the hyperplastic background may be attributed to secondary, additional, stochastic mutation related to neural or humoral factors, resulting in the transition from hyperplasia to neoplasia [5,6].

It is notable that the P86S mutant GCGR is predominantly localized in the endoplastic reticulum, while the wild type of GCGR is found mainly on the plasma membrane. Compared to cells expressing the wild type of GCGR, cells expressing the P86S mutant GCGR show lower affinity to glucagon and produce less cyclic adenosine monophosphate (cAMP) and intracellular calcium after stimulation by physiological concentrations of glucagon [7,8].

9.5 Clinical features

Mahvash disease is characterized by the presence of high glucagon levels (hyperglucagonemia) without glucagonoma syndrome (e.g., necrolytic migratory erythema, reduced amino acids, and hyperglycemia), pancreatic α cell hyperplasia, and PNET (in 10 out of 11 confirmed cases) along with microadenomas, although fibrosis in the pancreatic head instead of PNET was observed in one case.

Common clinical symptoms of Mahvash disease include abdominal pain and discomfort without visual loss. Some patients with Mahvash disease may be asymptomatic apart from very high glucagon levels [1].

9.6 Diagnosis

PNET and remarkable hyperglucagonemia without glucagonoma syndrome are clinical hallmarks of Mahvash disease.

Therefore, diagnosis of Mahvash disease involves biochemical assessment of glucogon levels, radiographic imaging study of pancreatic tumors, histological detection of hypertrophic pancreas and PNET as well as pancreatic ademoma, and molecular identification of *GCGR* mutation [1]. The key macroscopic, microscopic, and immunohistochemical characteristics of PNET are provided in Chapter 3.

Differential diagnoses of Mahvash disease include hyperglucagonemia, pancreatic hypertrophy, and possibly hypoglycemia.

9.7 Treatment

Due to the absence of functional GCGR, PNET in Mahvash disease often behaves clinically as a nonfunctioning tumor. Therefore, tumors <2–3 cm in size may be monitored and those >3 cm should be surgically resected (see Chapter 3) [1].

Patients should undergo yearly measurement of glucagon levels and biannual imaging surveillance to detect any recurrent tumors at an early stage.

9.8 Prognosis

Given the relatively small number of confirmed cases reported so far (11 cases to be exact), prognosis for Mahvash disease is uncertain. Based on data available on pancreatic endocrine tumors (see Chapter 3), the overall 5-year survival rate for Mahvash disease (which tends to have nonfunctional PNET) is expected to be 30%.

With an autosomal recessive mode of inheritance, siblings of patients with Mahvash disease have a 25% probability of acquiring Mahvash disease. The probability for the offspring of patients with Mahvash disease to inherit the condition is also small [1].

References

1. Rhyu J, Yu R. Mahvash disease: An autosomal recessive hereditary pancreatic neuroendocrine tumor syndrome. *Int J Endocrine Oncol.* 2016;3(3):235–43.
2. Quesada I, Tudurí E, Tudurí C, Nadal A. Physiology of the pancreatic α-cell and glucagon secretion: Role in glucose homeostasis and diabetes. *J Endocrinol.* 2008;199:5–19.
3. Ouyang D, Dhall D, Yu R. Pathologic pancreatic endocrine cell hyperplasia. *World J Gastroenterol.* 2011;17(2):137–43.
4. Yu R, Dhall D, Nissen NN, Zhou C, Ren SG. Pancreatic neuroendocrine tumors in glucagon receptor-deficient mice. *PLoS One.* 2011;6(8):e23397.

5. Yu R, Ren SG, Mirocha J. Glucagon receptor is required for long-term survival: A natural history study of the Mahvash disease in a murine model. *Endocrinol Nutr.* 2012;59(9):523–30.
6. Yu R, Chen CR, Liu X, Kodra JT. Rescue of a pathogenic mutant human glucagon receptor by pharmacological chaperones. *J Mol Endocrinol.* 2012;49(2):69–78.
7. Ro C, Chai W, Yu VE, Yu R. Pancreatic neuroendocrine tumors: Biology, diagnosis, and treatment. *Chin J Canc.* 2013;32(6):312–24.
8. Lucas MB, Yu VE, Yu R. Mahvash disease: Pancreatic neuroendocrine tumor syndrome caused by inactivating glucagon receptor mutation. *J Mol Genet Med.* 2013;7:4.

10
McCune–Albright Syndrome

10.1 Definition

McCune–Albright syndrome (MAS) is a rare disease resulting from an early embryonic somatic mosaic activating mutation of the GNAS gene (guanine nucleotide binding protein, alpha stimulating, which encodes the cAMP pathway-associated G-protein, Gsα). Originally defined as the triad of polyostotic fibrous dysplasia of bone, café au-lait skin pigmentation, and precocious puberty, the clinical spectrum of MAS has since been expanded to include hyperthyroidism, Cushing's syndrome, pituitary gigantism/acromegaly (if mutated cells are present in thyroid, adrenal, and/or pituitary tissues), renal phosphate wasting with or without rickets/osteomalacia, and hepatic and cardiac involvement as part of the condition [1].

10.2 Biology

Mosaicism refers to the coexistence of genetically distinct cell populations in a single individual (that has developed from a single fertilized egg) as a result of post-zygotic genetic events. Mosaicism differs from chimerism, which involves two or more genotypes arising from the fusion of more than one fertilized zygote in the early stages of embryonic development.

Mosaicism occurring in somatic cells is known as somatic mosaicism. Somatic mosaic mutation may be clinically silent or may affect a portion of the body but will not pass to offspring. Whereas somatic mosaic mutations that take place early in development may cause widespread disease, those taking place late in development show limited disease. On the contrary, mosaicism occurring in germline cells is known as germline mosaicism. Although germline mosaicism does not have phenotypic consequences on an individual, it will pass to offspring.

Somatic mosaicism may be attributed to errors in the replication of segments of chromosomes or of whole chromosomes (chromosomal aneuploidy), copy-neutral reciprocal gains and losses (acquired uniparental disomy or loss of heterozygosity), changes in nuclear or mitochondrial DNA, an acquired point mutation, or spontaneous reversion of an existing DNA mutation [2].

MAS represents an outcome of mosaic somatic activating (gain of function) mutations in GNAS that produce amino acid substitution of either residue Arg201 or Gln227 in the α subunit of the stimulatory G protein (Gsα) associated with the cAMP-dependent pathway, leading to disruption of the signaling turn-off mechanism, autonomous hyperfunction, and tumorigenesis. Given the ubiquitous expression of Gsα, MAS can affect tissues derived from the ectoderm, mesoderm, and endoderm, including the skin, skeleton, and endocrine organs [3].

10.3 Epidemiology

MAS has an overall incidence of one to ten cases per million worldwide. Within the spectrum of MAS diseases, fibrous dysplasia (particularly the monostotic form) is estimated to account for up to 7% of all benign bone tumors. MAS affects both sexes, without a predilection for any particular population groups.

10.4 Pathogenesis

MAS is a noninherited disease caused by random, postzygotic, mosaic mutations in the *GNAS* gene, leading to some cells with a normal version of the *GNAS* gene and others with a mutated version. Therefore, the number and location of cells containing the mutated *GNAS* gene determine the features and severity of MAS.

Spanning over 20-kb on chromosome 20q13, the GNAS gene consists of 13 exons and encodes the ubiquitously expressed stimulatory α-subunit of the G protein (Gsα). The Gsα protein is 394 amino acids long with a mass of 46 kDa and contains two domains: a guanosine triphosphate hydrolase (GTPase) domain involved in the binding and hydrolysis of GTP and a helical domain determining the intrinsic GTPase activity of the α subunit. Gsα couples hormone receptors to adenyl cyclase for the generation of intracellular cAMP that then mediates Gsa-coupled hormone signaling.

GNAS mutations associated with MAS are predominated by the amino acid substitutions R201C, R201H, R201S, or R201G (>95%) and, rarely, Q227R or Q227K (<5%). These missense mutations produce a constitutively activated form of Gsa with reduced GTPase catalytic ability, making it impossible to control the Gsa activation and cAMP production. Enhanced Gsα signaling and cAMP generation accelerate the commitment of bone marrow stromal cells into the osteoblastic lineage and their further differentiation into osteoblasts, leading to the formation of fibrous dysplastic lesions consisting of fibrous cells that express early osteoblastic markers (e.g., alkaline phosphatase). Furthermore, increased Gsα signaling expedites the action of alpha-MSH for melanin production and skin lesion formation. Persistently high levels of cAMP also activate protein kinase A (PKA) and mitogen-activated protein kinase (MAPK, as seen in cardiac hypertrophy), induce overexpression of cellular oncogene *fos* (*c-fos* gene, as seen in fibrous dysplasia), and alter the processing of the FGF23 protein by GALNT3 and furin, leading to elevated levels of the inactive C-terminal fragment of FGF23 [3].

Besides somatic activating (gain-of-function) *GNAS* mutations in MAS, germline inactivating (loss-of-function) *GNAS* mutations are associated with multiple phenotypes. These include pseudopseudohypoparathyroidism (PPHP, due to an inactivating mutation of the paternal *GNAS* allele resulting in expression of the protein product Gsα only from the maternal allele), pseudohypoparathyroidism 1A (due to an inactivating mutation of the maternal *GNAS* allele resulting in expression of the protein product Gsα only from the paternal allele), pseudohypoparathyroidism IB (due to deletion of the regulatory differentially methylated region of the *GNAS* locus), and progressive osseous heteroplasia (POH; due possibly to an inactivating *GNAS* mutation of the paternal allele).

10.5 Clinical features

Clinically, MAS may present with a spectrum of symptoms involving the skin, bones, endocrine organs, and other tissues [4–6].

10.5.1 Skin

The first manifestation of MAS is usually café-au-lait spots/skin macules (typically asymmetric, with jagged, irregular borders, distributed along Blaschko's line without crossing the midline), appearing at or shortly after birth.

10.5.2 Bones

Fibrous dysplasia of bones is the most frequent finding in MAS, which often emerges in the first few years of life and expands during childhood, with no clinically significant bone lesions appearing after age 15 years. Fibrous dysplasia may present as an isolated, asymptomatic monostotic lesion (i.e., monostotic fibrous dysplasia involving a single bone) or as severe, painful, disabling polyostotic disease (i.e., polyostotic fibrous dysplasia affecting more than one bone), with progressive scoliosis, facial deformity, and loss of mobility, vision, and/or hearing. The affected bones often show expansive lesions with endosteal scalloping, thin cortex, and an intramedullary tissue matrix (so-called ground glass). In rare cases (<1%, often in association with previous radiation treatment), a sarcoma may develop within fibrous dysplasia.

10.5.3 Endocrine organs

Clinical symptoms of MAS involving endocrine organs range from precocious puberty, testicular abnormalities, thyroid disease, FGF23-mediated phosphate wasting, growth hormone excess, to hypercortisolism.

Gonadotropin-independent *precocious puberty* commonly occurs in girls (approximately 85%), with recurrent ovarian cysts leading to intermittent estrogen production, breast development, growth acceleration, and vaginal bleeding. Precocious puberty in boys (approximately 15%) shows autonomous testosterone production, with growth acceleration, pubic and axillary hair, acne, and aggressive and/or inappropriate sexual behavior.

Other symptoms include testicular lesions with or without associated gonadotropin-independent precocious puberty, thyroid lesions with or without non-autoimmune hyperthyroidism, growth hormone excess, FGF23-mediated phosphate wasting with or without hypophosphatemia in association with fibrous dysplasia, and neonatal hypercortisolism. The prognosis for individuals with fibrous dysplasia/MAS is based on disease location and severity.

Testicular abnormalities typically manifest as unilateral or bilateral macro-orchidism in males (approximately 85%), with discrete hyper- and hypoechoic lesions and microlithiasis, corresponding to areas of Leydig and/or Sertoli cell hyperplasia.

Thyroid disease manifests as hyperthyroidism in one-third of individuals with MAS due to increased hormone production and increased conversion of thyroxine (T4) to triiodothyronine (T3), with mixed cystic and solid lesions interspersed with areas of normal-appearing tissue.

FGF23-mediated phosphate wasting is characterized by increased production of phosphaturic hormone FGF23 in fibrous dysplasia tissue, leading to renal tubulopathy and phosphate wasting (hypophosphatemia), with risk of fractures, and fibrous dysplasia-related bone pain.

Growth hormone excess is seen in 10%–15% of individuals with MAS harboring *GNAS* mutations in the anterior pituitary, leading to autonomous growth hormone production, hyperprolactinemia, linear growth acceleration, acromegaly, expansion of craniofacial fibrous dysplasia, and macrocephaly.

Hypercortisolism occurs in infants as a result of excess cortisol production from the fetal adrenal gland.

10.5.4 Other tissues and organs

Hepatitis and hepatic adenomas may occur in infants. Gastroesophageal reflux and gastrointestinal polyps are observed in childhood. Pancreatitis and intraductal papillary mucinous neoplasms are also found. Benign intramuscular myxomas (Mazabraud syndrome) are also found incidentally. Cancers affecting the bone, thyroid, testicles, and breast are observed occasionally.

10.6 Diagnosis

A diagnosis of MAS relies on the finding of two or more typical clinical features. Individuals with only monostotic fibrous dysplasia require identification of a *GNAS* somatic activating mutation for confirmation, based on sequencing analysis of exons 8 and 9, with mutations p.Arg201His and p.Arg201Cys detected in 8%–90% and 75%–100% of cases, respectively.

10.7 Treatment

For fibrous dysplasia, no medical therapies are currently available to alter the disease course, and disease management should center on minimizing morbidity related to repairing fractures and preventing deformity (by orthopedic surgery). For precocious puberty, use of the aromatase inhibitor letrozole and/or the estrogen receptor modulator tamoxifen in girls prevents bone age advancement and compromise of adult height. For thyroid disease, thyroidectomy is useful for persistent hyperthyroidism, while methimazole is effective in most other cases related to hyperthyroidism. For growth hormone excess, use of octreotide and the growth hormone receptor antagonist pegvisomant (alone or in combination) is the preferred first-line treatment.

Patients with MAS should avoid contact sports and other high-risk activities with significant skeletal involvement, prophylactic optic nerve decompression (craniofacial fibrous dysplasia), surgical removal of ovarian cysts, and radiotherapy for fibrous dysplasia.

Surveillance for persons with MAS includes monitoring (i) infants for clinical signs of hypercortisolism, (ii) children for growth acceleration (related to precocious puberty and/or growth hormone excess), (iii) children age under

5 years for thyroid function, (iv) males for testicular lesions with physical examination and testicular ultrasound (v) existing fibrous dysplasia and development of new lesions, (vi) phosphorus levels for the development of hypophosphatemia, (vii) vision and hearing in craniofacial fibrous dysplasia, (viii) progressive scoliosis with periodic skull CT in spine fibrous dysplasia, and (ix) children with thyroid abnormalities by ultrasound examination and thyroid function test.

10.8 Prognosis

The prognosis for individuals with MAS varies with disease location and severity. Although patients with skin lesions require minimal medical attention, those with extensive bone disease need to be treated promptly to reduce potential sequelae such as loss of mobility, progressive scoliosis, facial deformity, and loss of vision and/or hearing.

As MAS is a sporadic, noninherited disorder, the risk for siblings is similar to that for the general population.

References

1. Boyce AM, Collins MT. Fibrous Dysplasia/McCune-Albright Syndrome. In: Pagon RA, Adam MP, Ardinger HH, et al, editors. *GeneReviews®* [Internet]. Seattle, WA: University of Washington, Seattle; 1993–2017. 2015.
2. Freed D, Stevens EL, Pevsner J. Somatic mosaicism in the human genome. *Genes (Basel).* 2014;5(4):1064–94.
3. Salpea P, Stratakis CA. Carney complex and McCune Albright syndrome: An overview of clinical manifestations and human molecular genetics. *Mol Cell Endocrinol.* 2014;386(1–2):85–91.
4. Salenave S, Boyce AM, Collins MT, Chanson P. Acromegaly and McCune-Albright syndrome. *J Clin Endocrinol Metab.* 2014;99(6):1955–69.
5. Turan S, Bastepe M. GNAS Spectrum of disorders. *Curr Osteoporos Rep.* 2015;13(3):146–58.
6. *Atlas of genetics and cytogenetics in oncology and haematology.* GNAS (GNAS complex locus). http://atlasgeneticsoncology.org/Genes/GC_GNAS.html. Accessed on May 15, 2017.

11 Multiple Endocrine Neoplasia

11.1 Definition

Consisting of several distinct autosomal dominant disorders that affect the hormone-producing glands (i.e., the endocrine system), multiple endocrine neoplasia (MEN) is characterized by the formation of tumors in two or more endocrine glands within a single patient and by the presence of specific gene mutations.

Based on the specific genes involved, types of hormones made, and characteristic signs and symptoms, MEN is distinguished into MEN type 1 (MEN1), MEN type 2A (MEN2A), MEN type 2B (MEN2B), and MEN type 4 (MEN4).

MEN1 harbors mutations in the tumor suppressor gene *MEN1* (encoding menin) and often manifests as parathyroid, anterior pituitary, and pancreatic tumors. Of the two MEN1 variants identified, MEN1 Burin shows a high incidence of prolactinoma (40%) and a low incidence of gastrinoma (10%), while familial isolated hyperparathyroidism has a tendency to develop hyperparathyroidism without other endocrinopathies [1,2].

MEN2A is associated with mutations in the rearranged during transfection (*RET*) proto-oncogene encoding a tyrosine kinase receptor. Accounting for 55% of MEN2 cases, MEN2A presents with medullary thyroid carcinoma (MTC), pheochromocytoma, and parathyroid tumors. Being a subtype (variant) of MEN2A, familial medullary thyroid carcinoma (FMTC) makes up 35% of MEN2 cases and has MTC as its only clinical manifestation. However, FMTC demonstrates a later onset age and a lower penetrance other MEN2 types [3,4].

MEN2B also contains mutations in the *RET* gene. Representing 5%–10% of MEN2 cases, MEN2B typically displays MTC and pheochromocytoma [3,4].

MEN4 (also called MENX) is attributed to germline mutations in the cyclin-dependent kinase inhibitor gene (*cdkn1b*) encoding the p27 protein. Apart from parathyroid and anterior pituitary tumors, MEN4 may also cause tumors of the adrenals, kidneys, and reproductive organs [5–7].

11.2 Biology

The endocrine system comprises a network of hormone-producing glands (e.g., adrenal, pituitary, parathyroid, thyroid, and pancreatic endocrine) that secrete various types of hormones in order to regulate the function of cells and tissues throughout the body.

MEN represents a group of autosomal dominant tumor predisposition syndromes that target the endocrine system. Typically showing tumors (neoplasia) in at least two endocrine glands as well as other organs and tissues, MEN is characterized by the occurrence of specific mutations in relevant genes.

Although one copy of the altered *RET* or *cdkn1b* gene is sufficient for MEN2 or MEN4 to cause the disorder, two copies of the altered *MEN1* gene are required for MEN1 to trigger tumor formation, with one mutated gene inherited and a mutation in the second copy of the *MEN1* gene occurring in a small number of cells during a person's lifetime ("two-hit" model).

In most MEN1 cases, one mutated copy of the *MEN1* gene is inherited from an affected parent. However, in some MEN1 patients who have no affected relatives, mutation in the *MEN1* gene may arise sporadically. Further, in about 10% of MEN1 kindreds, an identifiable *MEN1* mutation is lacking [1,2].

MEN2A (also known as Sipple syndrome) and MEN2B (sometimes called MEN-3 or mucosal neuroma syndrome, formerly Wagenmann–Froboese syndrome) both contain germline mutations in the *RET* proto-oncogene and develop multicentric, bilateral MTC (a cancer of the parafollicular calcitonin secreting C cells) and bilateral pheochromocytoma in 50% of cases. As a variant of MEN2A, FMTC presents with MTC only [3,4].

MEN4 covers some MEN1-like cases that do not have mutations in the *MEN1* gene. Through comparison of germline mutation in the *cdkn1b* gene from rats with multiple endocrine tumors and that in a human homolog, the role of the *cdkn1b* gene and its encoded protein (p27) in the tumorigenesis of some MEN1-like cases with intact *MEN1* (i.e., MEN4 or MENX) was confirmed [5–7].

11.3 Epidemiology

MEN1 has an estimated prevalence of 2–20 cases per 100,000. Inherited in an autosomal dominant pattern, MEN1 shows a high degree of penetrance, with biochemical signs detected in >95% and clinical signs in 80% of patients by the fifth decade of life.

MEN2 affects one per 200,000 live births. Among the MEN2 subtypes, MEN2A, FMTC, and MEN2B account for 70%–80%, 10%–20%, and 5% of all MEN2 cases, respectively.

MEN4 appears to be rare. Of the 10% *MEN1* mutation-free patients fulfilling the diagnostic criteria for MEN1, a small number (up to 3%) contain mutations in the *CDKN1B* gene and belong to MEN4.

11.4 Pathogenesis

The *MEN1* gene is a tumor suppressor gene located on the long arm of chromosome 11 (11q13). About 10 kb in length and containing 10 exons, the *MEN1* gene encodes a 610-amino acid protein called menin. Through interaction with transcription factors including activating protein-1 (AP-1), JunD, nuclear factor-κB (NF-κB), β-catenin, mothers against decapentaplegic (SMAD) family members, and estrogen receptor α (ERα), menin plays an important role in cell proliferation, apoptosis, and genome integrity. Mutations inactivating both copies of the *MEN1* gene result in a malfunctional menin, which loses its capacity to control cell division. To date, more than 300 different *MEN-1* germline mutations (non-sense deletions in 80% cases, in-frame deletion in shift mutations in familial isolated hyperparathyroidism, loss of heterozygosity in normal chromosome of the unaffected parent) have been identified [1,2].

The *RET* proto-oncogene related to MEN2A, MEN2B, and FMTC is located at the centrometric region of chromosome 10 (10q11-2). This gene spans more than 60 kb with 21 exons and encodes a transmembrane tyrosine kinase with a long extracellular domain, a single transmembrane region, and two cytoplasmic tyrosine kinase domains. This protein has a critical function in cell migration and development. Mutations (e.g., missense mutation at codons 609, 611, 618, 620, or 634; point mutation at codon 918) in the *RET* gene overactivates the protein's signaling function, which in turn triggers cell growth and division in the absence of signals from outside the cell. Besides *RET* gain of function mutations in MEN2, germline *RET* loss-of-function pathogenic variants are associated with Hirschsprung disease (HSCR), which typically causes enlargement of the bowel and constipation or obstipation in neonates [3,4].

The *CDKN1B* tumor suppressor gene in MEN4 is located on chromosome 12p13 and encodes a 196 aa protein (p27) belonging to the kinase inhibitory protein/cyclin-dependent kinase (CDK) inhibitor interacting protein (KIP/CIP) family of cell cycle inhibitors; p27 binds to cyclinE/CDK2 and cyclinA/CDK2 complexes in response to either mitogenic or anti-mitogenic stimuli, prevents pRb phosphorylation, and stops the cells in the G1 phase from progression to the S phase. Heterozygous loss-of-function mutations (eight identified so far) in the CDKN1B gene reduce the amount of functional p27, allowing cells to grow and divide unchecked. Interestingly, germline *CDNK1B* mutations may occasionally occur in sporadic (i.e., nonfamilial) cases of primary hyperparathyroidism [5,7].

11.5 Clinical features

MEN1 is linked to more than 20 endocrine and non-endocrine tumors. Parathyroid tumors cause hypercalcemia, lethargy, depression, confusion, anorexia, constipation, nausea, vomiting, diuresis, dehydration, hypercalciuria, kidney stones, increased bone resorption/fracture risk, hypertension, and shortened QT interval. Prolactinoma, affecting the pituitary gland, manifests as oligomenorrhea/amenorrhea and galactorrhea in females and sexual dysfunction in males. An adrenocortical tumor may be associated with primary hypercortisolism or hyperaldosteronism [1,2].

MEN2 is responsible for MTC, pheochromocytoma, or parathyroid tumors. Although MTC usually behaves in a relatively indolent fashion, some MEN2A patients may have aggressive MTC, similar to MEN2B-related MTC. Pheochromocytoma in MEN2A and MEN2B is often bilateral and multiple, with palpitations, headaches, and anxiety as the main symptoms [3,4].

MEN4 is similar to MEN1, with a high incidence of pituitary tumors and primary hyperparathyroidism [7].

11.6 Diagnosis

A combination of family history, clinical features, biochemical tests, and genetic detection is utilized for MEN diagnosis. A biochemical screening program typically includes measurements of intact PTH, serum calcium, prolactin, somatomedin C, glucose, insulin, pro-insulin, gastrin, pancreatic polypeptide, glucagon, and ingestion of a test meal followed by measurement of pancreatic polypeptide and gastrin.

Diagnosis of MEN1 is based on any one of three criteria: (i) an individual with a known MEN1 gene mutation but does not have clinical or biochemical evidence of disease, (ii) an individual with one MEN1-associated tumor and a first-degree relative diagnosed with MEN1, and (iii) an individual with at least two MEN1-associated tumors [1,2].

Diagnosis of MEN2 requires fulfilling clinical criteria. MEN2A shows two or more specific endocrine tumors (MTC, pheochromocytoma, or parathyroid adenoma/hyperplasia) in a single individual or in close relatives. FMTC shows four or more cases of MTC in the absence of pheochromocytoma or parathyroid adenoma/hyperplasia in families. MEN2B shows early-onset MTC, mucosal neuromas of the lips and tongue, as well as medullated corneal nerve fibers, distinctive facies with enlarged lips, and an asthenic "marfanoid" body habitus [3,4].

Molecular identification of a heterozygous germline *RET* pathogenic variant is indicative of primary C-cell hyperplasia or MTC or a clinical diagnosis of MEN2. Molecular identification of a heterozygous germline *RET* pathogenic variant establishes the diagnosis if clinical features are inconclusive [3,4].

Diagnosis of MEN4 should center on molecular identification of *CDKN1B* pathogenic variants in a MEN1-like case with an intact *MEN1* gene [7].

11.7 Treatment

Treatment options for MEN include surgery, radiotherapy, and chemotherapy [4]. Surgical removal of the thyroid gland and lymph node dissection are recommended for MTC. External beam radiation therapy (EBRT) or intensity-modulated radiation therapy (IMRT) may be considered for advanced locoregional disease. Chemotherapy (e.g., kinase inhibitors) has some value for metastatic MTC.

Adrenalectomy is a treatment of choice for pheochromocytoma uncovered by biochemical testing and radionuclide imaging. It should be noted that dopamine D_2 receptor antagonists and β-adrenergic receptor antagonists present a high risk for adverse reactions in individuals with pheochromocytoma.

Surgery to remove one or more parathyroid glands is applicable to primary hyperparathyroidism; and medications to reduce parathyroid hormone secretion may also be administered.

Prophylactic thyroidectomy may be considered for individuals with an identified germline *RET* pathogenic variant.

Surveillance for MEN includes annual measurement of serum calcitonin concentration to detect residual or recurrent MTC after thyroidectomy monitoring for possible hypoparathyroidism after thyroidectomy and parathyroid autotransplantation and annual biochemical screening for patients with a germline *RET* pathogenic variant and with initial negative screening results for pheochromocytoma.

11.8 Prognosis

MEN1 as well as MEN4 patients have a generally good prognosis, with a 15-year survival rate of 93%. Glucagonoma, VIPoma, somatostatinoma, and nonfunctioning pancreatic endocrine tumors in MEN1 are associated with a threefold to fourfold increased risk of death.

MEN2A patients with MTC have 5- and 10-year survival rates of 90% and 75%, respectively. MEN2B patients with clinically apparent, often aggressive MTC have a poorer prognosis than MEN2A patients.

References

1. Giusti F, Marini F, Brandi ML. Multiple endocrine neoplasia type 1. In: Pagon RA, Adam MP, Ardinger HH, et al, editors. *GeneReviews®* [Internet]. Seattle, WA: University of Washington, Seattle; 1993–2017. 2005 [updated 2015 Feb 12].
2. Falchetti A. Genetics of multiple endocrine neoplasia type 1 syndrome: What's new and what's old. *F1000Res*. 2017;6:pii:F1000 Faculty Rev-73.
3. Marquard J, Eng C. Multiple endocrine neoplasia type 2. In: Pagon RA, Adam MP, Ardinger HH, et al, editors. *GeneReviews®* [Internet]. Seattle, WA: University of Washington, Seattle; 1993–2017. 1999 [updated 2015 Jun 25].
4. Norton JA, Krampitz G, Jensen RT. Multiple endocrine neoplasia: Genetics and clinical management. *Surg Oncol Clin N Am*. 2015;24(4):795–832.
5. Pellegata NS. MENX and MEN4. *Clinics (Sao Paulo)*. 2012;67 Suppl 1:13–8.
6. Thakker RV. Multiple endocrine neoplasia type 1 (MEN1) and type 4 (MEN4). *Mol Cell Endocrinol*. 2014;386(1–2):2–15.
7. Schernthaner-Reiter MH, Trivellin G, Stratakis CA. MEN1, MEN4, and Carney complex: Pathology and molecular Genetics. *Neuroendocrinology*. 2016;103(1):18–31.

12 Von Hippel–Lindau Syndrome

12.1 Definition

Von Hippel–Lindau syndrome (VHL syndrome) is a rare, autosomal, dominantly inherited disorder resulting from mutations of the *VHL* gene on chromosome 3. Patients with VHL syndrome show increased risk for tumors and cysts (fluid-filled sacs) in various parts of the body, including retinal and central nervous system hemangioblastomas, renal cysts and clear cell renal cell carcinoma (RCC), pheochromocytoma in the adrenal glands, pancreatic cysts and neuroendocrine tumors, endolymphatic sac tumors in the inner ear, epididymal and broad ligament cysts [1].

12.2 Biology

An early description of VHL syndrome stems from the reports by Collins in 1894 and von Hippel in 1904 concerning the cases of familial retinal hemangioblastomas (then called "angiomatosis retinae"). Visceral features of VHL syndrome (including renal cysts and tumors as well as epididymal cysts) were subsequently documented by Brandt in 1921. After a careful review of clinical and pathologic features of 40 patients (e.g., cysts and tumors in the medulla and spinal cord in association with retinal hemangioblastomas and visceral tumors), the term "central nervous system angiomatosis" was proposed by Lindau for this group of diseases in 1927. The eponymous term "von Hippel–Lindau disease" was finally adopted in honor of von Hippel (for his description of retinal hemangioblastomas or von Hippel tumors) and Lindau (for cerebellar hemangioblastomas or Lindau's tumors) [1,2].

The *VHL* gene responsible for VHL syndrome was located on the short arm of chromosome 3 (3p25–26) by Seizinger et al. in 1988 and later isolated and sequenced by Latif et al. in 1993. It is now clear that germline mutations of the *VHL* gene inactivate the VHL protein (pVHL) and increase production of vascular endothelial growth factor (VEGF), platelet Derived Growth Factors (PDGF), transforming growth factor alpha (TGF-α). These changes render the patients with VHL syndrome susceptible to multiple benign and malignant tumors, as well as to cysts in multiple organs, including the cerebellum, brainstem, spinal cord, retina, endolymphatic sac, kidneys, adrenal glands, pancreas, broad ligament (female), and epididymis (male) [1,2].

Although most autosomal dominant conditions require only one altered copy of a gene in each cell to cause the disorder, VHL syndrome needs two altered copies of the *VHL* gene to trigger tumor and cyst formation. The penetrance of VHL syndrome is high as individuals with an inherited *VHL* mutation will eventually acquire a mutation in the second copy of the gene in some cells and become symptomatic by 70 years of age. However, the *de novo* manifestation that occurs during the formation of reproductive cells (eggs or sperm) or very early in development is noted in up to 20% of newly diagnosed of VHL cases [1,2].

12.3 Epidemiology

VHL syndrome has an estimated incidence of 1 case per 36,000 live births, with first manifestations emerging in the second decade of life and complete penetrance by 70 years of age. Notably, an endolymphatic sac tumor has a mean age of onset at 22 years and occurs in 10% of VHL patients; retinal hemangioblastoma, 25 years and 60%; pheochromocytoma, 30 years and 20%; cerebellar and spinal hemangioblastomas, 33 years and 65%; pancreatic cysts, microcystic serous adenoma, and neuroendocrine tumor, 36 years and 35%–70%; renal clear cell carcinoma and cysts, 39 years and 45%; and epididymal and broad ligament cystadenomas, unknown age and >50% males.

12.4 Pathogenesis

Spanning over 10 kb on chromosome 3p25–26, the *VHL* gene is a highly conserved tumor suppressor gene consisting of three exons. It encodes a 213 amino acid protein (pVHL) with a molecular weight of 30 kDa (which is a glycan-anchored membrane protein found in both nuclear and cytoplasmic compartments) and a second, smaller 159 amino acid isoform with a molecular weight of 19 kDa as a result of alternate translation initiation from codon 54.

Conforming to the "two-hit" model of hereditary cancer, a germline VHL mutation resulting in a defective allele in all cells of the body (first hit) is followed by a second, somatic event (second hit, frequently allelic loss/loss of heterozygosity, 49%, or hypermethylation, 35%) leading to the tumorigenesis. To date, more than 500 germline pathogenic variants (due to missense mutations, 52%; frameshift, 13%; nonsense mutations, 11%; large/complete deletions, 11%; splice site variants, 7%; in-frame deletions/insertions, 6%; occurring mainly between codon 54 and the carboxy terminal [with hot spot codon 167]; affecting both isoforms of protein) responsible for the development of VHL syndrome have been identified.

The amino acid protein pVHL is involved in the targeting of hypoxia inducible factor (HIF) for proteasomal degradation, the regulation of apoptosis (p53 inactivation and increased NF-κB activity), and stabilization of microtubules and regulation of extracellular matrix; pVHL inactivation increases expression and stabilization of HIF proteins, leading to differential upregulation of genes for growth factors (TGF, PDGF), angiogenesis (VEGF), glucose metabolism (GLUT1, PFK1), and tumor microenvironment (LOX, MMP1).

Based on clinical phenotypes and genetic alterations, VHL syndrome can be subclassified into types 1, 2A, 2B, or 2C. VHL type 1 is associated with deletions, nonsense mutations, and other microdeletions/insertions leading to disruption of pVHL folding and activity and has a high risk for retinal angioma, central nervous system (CNS) hemangioblastoma, RCC, pancreatic cysts and neuroendocrine tumors but a low risk for pheochromocytoma. VHL type 2 is linked to missense mutations resulting in a substitution of an amino acid on the surface of the protein and shows a high risk for pheochromocytoma. Specifically, VHL type 2A contains a distinct missense mutation (p.Tyr169His) and has a high risk for pheochromocytoma, retinal angiomas, and CNS hemangioblastoma but a low risk for RCC. VHL type 2B harbors missense mutations of codon 167 (e.g., p.Arg167Gln or p.Arg167Trp) and has a high risk for pheochromocytoma (82% at 50 years), RCC (60% at 60 years), retinal angioma, CNS hemangioblastoma, pancreatic cysts, and neuroendocrine tumors. VHL type 2C has specific missense mutations at codons 238 and 259 and causes pheochromocytomas only.

Interestingly, homozygous or compound heterozygous pathogenic variants in *VHL* (e.g., p.Arg200Trp) are associated with familial erythrocytosis type 2 (ECYT2), which is characterized by increased circulating red blood cell mass, increased serum levels of erythropoietin, and normal oxygen affinity, without VHL-related tumors [3].

12.5 Clinical features

Clinical presentations of VHL syndrome are highly variable, depending on the type and location of the tumors and cysts formed.

The prototypic lesion of VHL syndrome is a CNS hemangioblastoma, which occurs in up to 72% of patients and often appears as a distinct, red, vascular mass (as small as 2 mm) with a thin layer of capsule in the cerebellum (16%–69%), brainstem (5%–22%), spinal cord (13%–53%), cauda equina (11%), or supratentorial location (1%–7%). Due to the mass effect on neighboring neural structures, patients often have headaches, vomiting, pain, sensory and motor loss, gait disturbances, or ataxia.

Retinal hemangioblastoma (or retinal angioma) occurs in 49%–62% of patients and is often the first manifestation of VHL. It is located in the periphery (50%) or at the optic nerve (50%) and may cause visual field defect or blindness (15%) as a result of retinal detachment, exudation, or hemorrhage.

Multiple renal cysts and clear cell subtype RCC occur in 60% and 30% of VHL patients, respectively, with lesions ranging from simple cysts to entirely solid lesions. Simple renal cysts tend to be asymptomatic, but complex cysts may progress to solid RCC masses. Large RCC may manifest as a renal mass with flank pain, hematuria, or a flank mass.

Pheochromocytoma arises in the adrenal medulla and occurs in up to 16% of patients. It can be bilateral and occasionally multifocal and produces excess norepinephrine, leading to paroxysmal or sustained hypertension, palpitations, tachycardia, headaches, sweating, pallor, and nausea. Paraganglioma is a pheochromocytoma in extra-adrenal locations, mostly along the sympathetic axis in the abdomen or thorax, and is often nonfunctional.

Pancreatic cysts rarely cause endocrine or exocrine insufficiency, although they may compress intestines or the bile duct. Pancreatic neuroendocrine tumors occur in 15%–56% of patients and are usually slow growing and hormonally inactive.

Endolymphatic sac tumors arise from the endolymphatic epithelium within the vestibular aqueduct, affect 10%–16% of patients, and may present as sudden, uni-, or bilateral hearing loss (due to an intralabyrinthine hemorrhage, endolymphatic hydrops, or otic capsule invasion), vertigo, tinnitus, and facial paresis.

Epididymal or papillary cystadenoma in males is common and typically asymptomatic, although it may lead to infertility (in bilateral cases). Papillary cystadenoma of the broad ligament in women is rare and usually asymptomatic.

12.6 Diagnosis

Diagnosis of VHL syndrome is based on clinical diagnostic criteria and molecular identification of a heterozygous germline *VHL* pathogenic variant (in case the clinical features are inconclusive) [4].

Current clinical diagnostic criteria include the finding of (i) CNS hemangioblastoma (including retinal hemangioblastoma); (ii) endolymphatic sac tumors (ELSTs); (iii) RCC, (iv) pheochromocytoma, paraganglioma, and/or glomus tumors; and (v) neuroendocrine neoplasm and/or multiple cysts of the pancreas.

An individual is considered as having VLH syndrome if presenting with: (i) at least two CNS hemangioblastomas, (ii) at least one CNS hemangioblastoma and one other manifestation described above or (iii) at least one of the manifestations described above and a pathogenic mutation in *VHL* gene or a first-degree relative with VHL.

12.7 Treatment

Microsurgical resection is the treatment of choice for CNS hemangioblastoma (especially large or symptomatic tumors). Stereotactic radiation or craniospinal radiation may be considered for patients with inoperable tumors. Retinal hemangioblastoma may be treated by laser photocoagulation, cryotherapy, diathermy, and xenon and external beam radiotherapy (when standard therapy has not prevented progression).

Renal cell carcinomas (>3 cm) should undergo early surgery (nephron-sparing or partial nephrectomy) or radiofrequency ablation. Bilateral nephrectomy is followed by renal transplantation.

Pheochromocytoma and paraganglioma are treated by laparoscopic surgery or partial adrenalectomy. Preoperative treatment with alpha-adrenergic blockade and an optional additional beta-adrenergic blockade may be considered.

Asymptomatic pancreatic cysts do not require treatment, although cysts causing obstructive symptoms may undergo laparoscopic surgical decompression. Pancreatic neuroendocrine tumors with high risk of metastases (≥3 cm, pathogenic variant in exon 3, with a doubling rate of less than 500 days) should be excised.

Endolymphatic sac tumors may be removed to relieve audiovestibular symptoms and to preserve hearing and vestibular function. Epididymal or broad ligament papillary cyst adenomas generally require no surgery unless they become symptomatic or fertility-threatening.

Drugs under development include an intravitreal VEGF receptor inhibitor (ranibizumab and bevacizumab) for retinal hemangioblastoma, a tyrosine kinase inhibitor (sunitinib) for pheochromocytoma and clear cell RCC, and thalidomide for multifocal spinal hemangioblastoma [1,2].

12.8 Prognosis

VHL patients have a life expectancy of 59.4 years in males and 48.4 years in females. The major cause of mortality is CNS hemangioblastoma.

Tobacco, chemicals, and industrial toxins that are known to affect VHL-involved organs should be avoided; contact sports should be avoided if adrenal or pancreatic lesions are present.

About 80% of individuals with VHL syndrome have an affected parent, and about 20% contain a *de novo* pathogenic variant (e.g., CCND1 or cyclin D1). The offspring of an individual with VHL syndrome are at a 50% risk of inheriting the *VHL* pathogenic variant. Prenatal testing may be considered if the pathogenic variant is identified in a family member [1,2].

References

1. Frantzen C, Klasson TD, Links TP, Giles RH. Von Hippel-Lindau Syndrome. In: Pagon RA, Adam MP, Ardinger HH, et al, editors. *GeneReviews®* [Internet]. Seattle, WA: University of Washington, Seattle; 1993–2017. 2000 [updated 2015 Aug 6].
2. Gläsker S, Neumann HPH, Koch CA, Vortmeyer AO. Von Hippel-Lindau Disease. In: De Groot LJ, Chrousos G, Dungan K, et al, editors. *Endotext* [Internet]. South Dartmouth, MA: MDText.com, Inc.; 2000–2015.
3. Chittiboina P, Lonser RR. Von Hippel-Lindau disease. *Handb Clin Neurol.* 2015;132:139–56.
4. Ben-Skowronek I, Kozaczuk S. Von Hippel-Lindau Syndrome. *Horm Res Paediatr.* 2015;84(3):145–52.

SECTION II
Hematopoietic and Lymphoreticular Systems

13 Myeloproliferative Neoplasms (MPN)

13.1 Definition

Tumors of the hematopoietic and lymphoreticular systems consist of two categories: (i) myeloid neoplasms and acute leukemia (which represent about 35% of all hematopoietic and lymphoreticular neoplasms) and (ii) mature lymphoid, histiocytic, and dendritic neoplasms (which include myeloma and represent 65% of all hematopoietic and lymphoreticular neoplasms) [1].

Myeloid neoplasms and acute leukemias encompass (i) myeloproliferative neoplasms (MPN), (ii) mastocytosis, (iii) myeloid/lymphoid neoplasms with eosinophilia and rearrangements, (iv) myelodisplastic/myeloproliferative neoplasms (MDS/MPN), (v) myelodisplastic syndrome (MDS), (vi) acute myeloid leukemia (AML), (vii) blastic plasmacytoid dendritic cell neoplasms (BPDCN), (viii) acute leukemia of ambiguous lineage (ALAL), (ix) B-lymphoblastic leukemia/lymphoma, and (x) T-lymphoblastic leukemia/lymphoma [2].

MPN (formerly myeloproliferative disorders or MPD) are a group of diseases characterized by the overproduction of one or more blood cell types (red cells, white cells, or platelets) by the bone marrow. MPN may be further separated into nine subgroups: (i) chronic myeloid leukemia (CML; *BCR-ABL1*+ gene), (ii) chronic neutrophilic leukemia (CNL), (iii) polycythemia vera (PV), (iv) primary myelofibrosis (PMF), (v) PMF—prefibrotic/early stage, (vi) PMF—overt fibrotic stage, (vii) essential thrombocythemia (ET), (viii) chronic eosinophilic leukemia not otherwise specified (CEL-NOS), and (ix) MPN—unclassifiable (see Table 13.1) [2].

It is of interest to note that blood cancers represent the fourth most common type of neoplasms in Canada in the year of 2016 (11%, after cancers of the lung 14%, breast 13%, colorectum 13%; but ahead of prostate cancer 10.6%). Among 22,340 newly diagnosed Canadian cases of blood cancers in 2016, lymphoma makes up 9,000 (40%), leukemia 5,900 (27%), myelodysplastic syndrome 3,850 (17%), myeloma 2,700 (12%), polycythemia vera (PV) 400 (1.8%), essential (primary) thrombocythemia (ET/PT) 290 (1.3%), and myelofibrosis 200 (0.9%). Further, between 1992–2008, the

Table 13.1 Key Characteristics of MPN Subtypes

MPN Subtype	Cellular Features	Genetic Mutation	Major Diagnostic Criteria
Chronic myeloid leukemia (CML)—*BCR-ABL1+*	Overproduction of granulocytes	BCR-ABL1 fusion	Persistent or increasing WBC (>10 × 10^9/L) and splenomegaly; persistent thrombocytosis (>1,000 × 10^9/L) and thrombocytopenia (<100 × 10^9/L); 20% or more basophils in the PB; 10%–19% blasts in the PB and/or BM; unresponsive to TKI therapy
Chronic neutrophilic leukemia (CNL)	Overproduction of neutrophils	CSF3R (T618I) no BCR-ABL1 fusion no JAK2 (V617F)	PB WBC ≥25 × 10^9/L; hypercellular BM; not meeting WHO criteria for BCR-ABL1+ CML, PV, ET, or PMF; no rearrangement of PDGFRA, PDGFRB, or FGFR1, or PCM1-JAK2; presence of CSF3R T618I
Polycythemia vera (PV)	Overproduction of RBC, accompanied by increased proliferation of WBC and platelets	JAK2 (V617F, 95%) no BCR-ABL1 fusion	Hemoglobin >16.5 g/dL in men, >16.0 g/dL in women; hematocrit >49% in men, >48% in women; or increased red cell mass; BM biopsy showing hypercellularity for age with trilineage growth (panmyelosis); presence of JAK2 or JAK2 exon 12 mutation
Primary myelofibrosis (PMF)	Lack of RBC and overproduction of WBC	JAK2 (V617F, 65%) no BCR-ABL1 fusion	Proliferation and atypia of megakaryocytes; reticulin and/or collagen fibrosis grades 2 or 3; not meeting WHO criteria for ET, PV, BCR-ABL1+ CML, myelodysplastic syndromes, or other myeloid neoplasm; presence of JAK2, CALR, or MPL mutation or of another clonal marker or absence of reactive myelofibrosis

(Continued)

Table 13.1 ***(Continued)*** **Key Characteristics of MPN Subtypes**

MPN Subtype	Cellular Features	Genetic Mutation	Major Diagnostic Criteria
Essential thrombocythemia (ET)	Overproduction of platelets	JAK2 (V617F, 55%) no BCR-ABL1 fusion	Platelet ≥450 × 10^9/L; BM biopsy showing proliferation mainly of the megakaryocyte lineage; not meeting WHO criteria for BCR-ABL1+ CML, PV, PMF, myelodysplastic syndromes, or other myeloid neoplasms; presence of JAK2, CALR, or MPL mutation
Chronic eosinophilic leukemia (CEL) NOS+	Overproduction of eosinophils	ETV6-FLT3 No PDGFRA, PDGFRB, FGFR1 rearrangements; no BCR-ABL1 fusion	Persistent PB eosinophils ≥1.5 × 10^9/L; either PB blasts >2%, BM blasts >5%,or abnormal cytogenetics(e.g.,trisomy 8 or isochromosome 17 or i[17q]); exclusion of other acute or chronic myeloid neoplasms (e.g., acute myeloid leukemia [AML], myelodysplastic syndrome [MDS], systemic mastocytosis [SM], and chronic myelomonocytic leukemia [CMML]).
MPN unclassifiable	Increased blasts and/or dysplasia	JAK2 (V617F, 20%) no BCR-ABL1 fusion	Early stages of PV, PMF, ET with no characteristic features; advanced stage MPN showing pronounced myelofibrosis, osteosclerosis, or transformation to a more aggressive stage; platelet ≥450 × 10^9/L or WBC ≥13 × 10^9/L; ≥1 hematopoietic cell line, <20% blasts in PB and BM

+CEL NOS excludes CEL with PDGFRA, PDGFRB, or FGFR1 rearrangements (see Chapter 15).

Abbreviations: BM, bone marrow; NOS, not otherwise specified; PB, peripheral blood; RBC, red blood cell; TKI, tyrosine kinase inhibitor; WBC, white blood cell; WHO, World Health Organization most common leukemias in Canada are chronic lymphocytic leukemia (CLL, 44%), acute myeloid leukemia (AML, 24%), chronic myeloid leukemia (CML, 12%), acute lymphocytic leukemia (ALL, 5%), and other leukemias (15%).

13.2 Biology

Forming the soft inner part of some bones (e.g., the skull, shoulder blades, ribs, pelvis, and backbones), the bone marrow (which consists of blood stem cells, mature blood-forming cells, fat cells, and supporting tissues) is largely responsible for producing blood cells after birth while the spleen does so during fetal life.

Lodged within the bone marrow, blood stem cells (also known as hematopoietic stem cells, HSC, or hematopoietic stem and progenitor cells, HSPC) are divided into myeloid and lymphoid stem cells. Myeloid stem cells give rise to myeloid lineage cells (i.e., red blood cells [erythrocytes], white blood cells [eosinophil, basophil, neutrophil, mast cell, and monocyte, but not lymphocyte], and platelets) in a process known as hematopoiesis (or hemapoiesis). Lymphoid stem cells differentiate into lymphoid lineage cells (T- and B-lymphocytes as well as natural killer cells) in a process known as lymphopoiesis (lymphocytopoiesis or lymphoid hematopoiesis) (Figure 13.1).

Red blood cells (RBC or erythrocytes) are involved in transportation of oxygen (via a carrier molecule called hemoglobin or Hgb) from the lungs to all other tissues in the body and in taking carbon dioxide back to the lungs for removal. When an insufficient number of RBC are present in the bloodstream (so-called anemia), the body will get an inadequate supply of oxygen, leading to tiredness, weakness, and shortness of breath.

White blood cells (WBC; including eosinophil, basophil, neutrophil, mast cell, monocyte/macrophage, and lymphocyte) are key components against infections in the host immune system. When an insufficient number of WBC are present in the bloodstream (so-called neutropenia), the body will have an increased susceptibility to microbial infection.

Collectively referred to as granulocytes, eosinophil, basophil, and neutrophil are derived from myeloblasts (a type of blood-forming cell in the bone marrow) and contain granules (formed by enzymes and other substances) of different sizes and colors. Among granulocytes, neutrophils are the most common type, and are essential in destroying invading bacteria in the blood. In addition, mast cell (or mastocyte) is a type of granulocyte derived from the myeloid stem cell. With many granules rich in histamine and heparin, mast cells have a similar appearance and function to basophils.

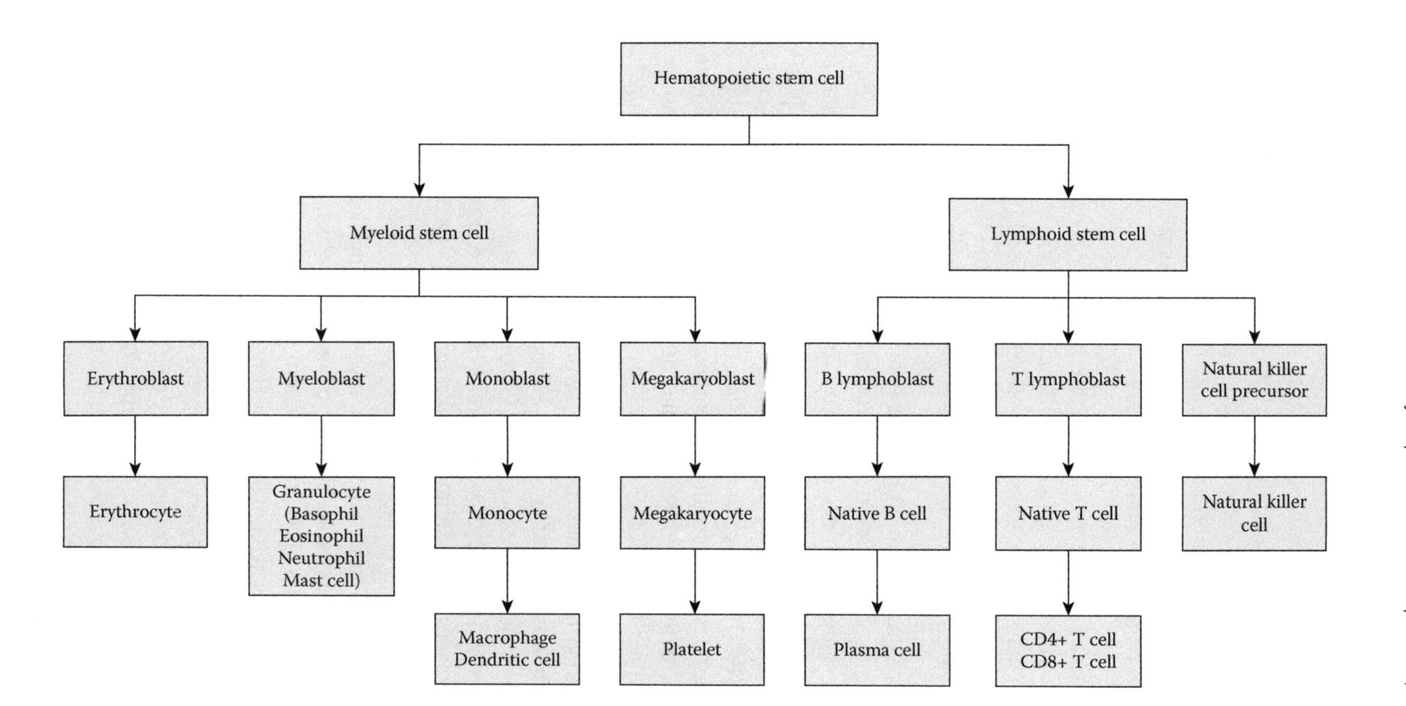

Figure 13.1 Differentiation of hematopoietic stem cell. Various stem cells, blast or precursor (immature) cells are located in the bone marrow, while mature cells are present in the peripheral blood.

Related to granulocytes, monocytes originate from blood-forming monoblasts (a type of blood stem cell in the bone marrow). They often enter body tissues and become macrophages after 1 day's circulation in the bloodstream and also dendritic cells (DC) upon moving to injured tissue. Monocytes may destroy certain microbes and help lymphocytes recognize and generate specific antibodies against microbes. DC specialize in antigen uptake, processing, and presentation, as well as stimulation of resting T cells in the primary immune response.

Platelets are cell fragments made by the megakaryocyte (a type of bone marrow cell) under the influence of thrombopoietin, and play an important role in plugging up holes in blood vessels caused by cuts or bruises. When an insufficient number of platelets is present in the bloodstream (so called thrombocytopenia), the body will bleed or bruise easily.

Lymphocytes evolve from lymphoblasts (a type of blood stem cell in the bone marrow) and represent the main cells in lymphoid tissue, which is distributed in the lymph nodes, thymus gland, spleen, tonsils, and adenoids, as well as the digestive and respiratory systems and bone marrow. Lymphocytes comprise two major populations: B lymphocytes (B cells) and T lymphocytes (T cells). Lymphocytes help regulate the immune system and protect the body from invading microbes. Specifically, T cells migrate to the thymus, differentiate further under the influence of thymic hormones, and have either helper (CD4) or cytotoxic (CD8) immunological functions (cellular immunity); B cells migrate directly to organs without undergoing modification in the thymus, become sensitized after exposure to antigen, and differentiate into plasma cells for antibody production (humoral immunity). Further, there is another class of large granular lymphocytes (distinct from T cells and B cells) called natural killer cells, which produce inflammatory cytokines and spontaneously kill malignant, infected, or "stressed" cells without prior sensitization.

Myeloid leukemias (also known as myelocytic, myelogenous, or nonlymphocytic leukemias) start in myeloid stem cells, which become myeloid-type white blood cells (granulocytes and monocytes but not lymphocytes), red blood cells, or platelet-making cells (megakaryocytes). In contrast, lymphoid leukemias (also known as lymphocytic or lymphoblastic leukemias) start in lymphoid stem cells, which become lymphoid-type white blood cells (B- and T-lymphocytes as well as NK cells). Lymphomas also evolve from lymphoid stem cells. However, in lymphocytic leukemia, cancer cells are mainly in the bone marrow and blood; in lymphoma, cancer cells are mostly in the lymph nodes and other tissues. Additionally, myeloma (or multiple myeloma) affects immunoglobulin-producing plasma cells derived from

B-lymphocytes and tends to accumulate in the bone marrow and on the surfaces of different bones in the body.

13.3 Epidemiology

CML is the most common MPN subtype, accounting for >10% of all new cases of leukemia. With an incidence of 1 case per 555, CML typically occurs in adults (mean age 64 years) and very rarely in children. A slight male predilection (male to female ratio of 1.3:1) is noted among CML patients.

CNL is a rare disease with fewer than 100 cases reported, has a relative incidence of 1–2 cases per 100 patients with BCR-ABL1-positive CML, and typically occurs in older adults.

PV has an incidence of 0.6–1.6 cases per million, with peak incidence among those 50–70 years old, and a slight male preponderance.

PMF has an annual incidence of 0.4–1.4 cases per 100,000 and a median age at diagnosis of 65 years. Although males are often affected in adult cases, girls are involved twice as frequently as boys in childhood cases.

ET has an incidence of 2.38 cases per 100,000 and a median age at diagnosis of 60 years (up to 20% patients under 40 years). Both sexes are equally affected among older patients, and more women than men are seen among younger patients.

CEL has an estimated incidence of 0.036 per 100,000. MPN unclassifiable represents about 2% of all MPN cases.

13.4 Pathogenesis

Although the tumorigenesis of MPN is not entirely clear, association between chronic MPN and constitutive activation of protein tyrosine kinases is often noted. Risk factors for CML are radiation exposure and old age. Molecularly, CML is linked to BCR-ABL fusion (t[9;22][q34;q11]). Other MPN subtypes do not contain BCR-ABL fusion but may harbor distinct genetic mutations of their own (Table 13.1).

13.5 Clinical features

CML and other MPN subtypes often present with increased or unexplained bleeding (nose, gums, and stool) or bruising, frequent or repeated infections, slow healing, pain or discomfort under the ribs on the left side, and excessives weating. Other symptoms include anemia, weakness, fatigue, weight loss, and enlarged spleen [3,4].

13.6 Diagnosis

Diagnosis of MPN involves laboratory assessment of RBC, WBC, platelet, and other cell levels in the peripheral blood (PB) and bone marrow (BM). Molecular identification of related gene mutations is also informative (Table 13.1).

CML is classified into chronic, accelerated, and blast phases based on the number of immature white blood cells (myeloblasts or blasts) in the PB or BM. Chronic phase CML typically has <10% blasts and shows fairly mild symptoms and good responses to treatment. Accelerated phase CML has 10%–20% blasts and at least 20% basophils and shows fever, poor appetite, and weight loss as well as poor responses to treatment. Blast phase (also called acute phase or blast crisis) CML has >20% blasts, displays fever, poor appetite, and weight loss [3].

PMF—prefibrotic/early stage and PMF—overt fibrotic stage are diagnosed using criteria similar to those for PMF (Table 13.1).

Differential diagnosis for CNL includes atypical CML (aCML). Although CNL has PB leukocytosis $\geq 25 \times 10^9$/L, aCML shows PB leukocytosis $\geq 13 \times 10^9$/L.

13.7 Treatment

For chronic phase CML, standard treatment consists of tyrosine kinase inhibitors (TKI) such as imatinib (Gleevec), nilotinib (Tasigna), or dasatinib (Sprycel). If one drug stops working (due to development of T315I mutation), the dose may be increased or another TKI may be used (e.g., bosutinib [Bosulif], ponatanib [Iclusig]) [4].

For CNL, treatments include oral hydroxyurea (to control leukocytosis and splenomegaly) and parenteral interferon α or its pegylated form (to control myeloproliferation); hypomethylating agents and ruxolitinib, thalidomide, cladribine, and imatinib provide palliative and transient benefits [4].

For PMF, treatment options range from ruxolitinib, hydroxyurea, thalidomide, lenalidomide, pomalidomide, cladribine, interferon-alpha, allogeneic peripheral stem cell or bone marrow transplantation, splenectomy, to radiotherapy (spleen and large lymph nodes) [4].

For ET, hydroxyurea is used to treat patients with high-risk essential thrombocythemia. Anagrelide represents a reasonable therapeutic alternative [4].

13.8 Prognosis

The 5-year overall survival rate for CML is 63.2%. The median survival of CNL patients is <2 years. Some CNL may progress into blast phase disease or chronic myelomonocytic leukemia (CMML). PMF has a median survival of 3.5–5.5 years, and patients <55 years of age have a median survival of 11 years. ET has a 10-year survival rate of 64%–80%, and transformation to acute myelogenous leukemia (AML) may occur in 0.6%–5% of cases [4].

References

1. Swerdlow SH, Campo E, Harris NL, et al. (eds). *WHO classification of tumours of haematopoietic and lymphoid tissues*. International Agency for Research on Cancer; World Health Organization. Lyon, France: IARC, 2008.
2. Arber DA, Orazi A, Hasserjian R, et al. The 2016 revision to the World Health Organization classification of myeloid neoplasms and acute leukemia. *Blood*. 2016;127(20):2391–405.
3. Granatowicz A, Piatek CI, Moschiano E, El-Hemaidi I, Armitage JD, Akhtari M. An overview and update of chronic myeloid leukemia for primary care physicians. *Kor J Fam Med*. 2015;36(5):197–202.
4. PDQ Adult Treatment Editorial Board. *Chronic Myeloproliferative Neoplasms Treatment (PDQ®): Health Professional Version*. PDQ Cancer Information Summaries [Internet]. Bethesda, MD: National Cancer Institute (US); 2002–2015.

14 Mastocytosis

14.1 Definition

Previously grouped among myeloproliferative neoplasms (MPN), mastocytosis represents a pathomorphologically and clinically heterogeneous spectrum of localized or systemic disorders characterized by the accumulation of neoplastic mast cells in the bone marrow and other organs/tissues. Mastocytosis consists of several variants, including cutaneous mastocytosis (CM), systemic mastocytosis (SM) (indolent systemic mastocytosis [ISM], smoldering systemic mastocytosis [SSM], systemic mastocytosis with an associated hematological neoplasm [SM-AHN], aggressive systemic mastocytosis [ASM], and mast cell leukemia [MCL]), mast cell sarcoma (MCS), and extracutaneous mastocytoma (Table 14.1) [1,2].

14.2 Biology

Derived from CD34+, c-kit+, CD13+ myeloid stem cells, mast cell (or mastocyte) is a type of granulocyte that contains granules rich in histamine and heparin and that demonstrates an appearance and function similar to basophil.

Mast cells are distributed in the skin, blood vessels, and nerves as well as in other tissues, and have the capacity to secrete various mediators (e.g., histamine, heparin, tryptase, chymase, platelet-activating factors, prostaglandin D2, LTC4/LTD4, proinflammatory cytokines, and chemokines) upon induction. Besides their role in allergies and anaphylaxis, mast cells are involved in wound healing, angiogenesis, immune tolerance, pathogen defense, and blood–brain barrier function.

Due to somatic mutations in the KIT gene (frequently D816V; occasionally V1815, D816F, and D820G), mast cells may show uncontrolled clonal proliferation and/or defective apoptosis, leading to mast cell clonal disorders, which can be divided into mastocytosis and monoclonal mast cell activation syndrome (MCAS), depending on the level of clonality.

As the most important mast cell clonal disorder, mastocytosis is characterized by the clonal proliferation of abnormal mast cells. In CM, neoplastic

Table 14.1 Key Characteristics of Mastocytosis Subtypes

Subtype	Cellular Features	Gene Mutation	Diagnostic Criteria
Cutaneous mastocytosis (CM)	Infiltrates of neoplastic mast cells in the dermis	KIT	Major diagnostic criteria: typical skin lesions or atypical lesions combined with a positive Dariers' sign (wheals and reddening produced by mechanical stroking of the lesions) and exclusion of other skin diseases. Minor diagnostic criteria: (i) >15 mast cells aggregating or >20 mast cells per high power field × (40) in skin biopsy; (ii) KIT D816V in lesional skin
Indolent systemic mastocytosis (ISM)	Less mast cell burden than other SM, 80% involving the skin	KIT	(i) SM criteria[a], (ii) absence of C-findings[b] and other clonal hematological diseases, (iii) <20% of mast cells in bone marrow sections
Smoldering systemic mastocytosis (SSM)	Intermediate between ISM and ASM	KIT	(i) SM criteria, (ii) absence of C-findings and other clonal hematological diseases but two or more B-findings[c]
Systemic mastocytosis with an associated hematological neoplasm (SM-AHN)	High mast cell burden, presence of clonal hematopoietic non-mast cell neoplasm	KIT	(i) SM criteria, (ii) demonstration of a clonal hematological non-mast cell disease
Aggressive systemic mastocytosis (ASM)	High mast cell burden, absence of MCL	KIT	(i) SM criteria, (ii) C-findings

(Continued)

Table 14.1 *(Continued)* Key Characteristics of Mastocytosis Subtypes

Subtype	Cellular Features	Gene Mutation	Diagnostic Criteria
Mast cell leukemia (MCL)	High mast cell burden, presence of chronic leukemia	KIT	(i) SM criteria, (ii) >20% of mast cells in bone marrow smears (or >10% in peripheral blood)
Mast cell sarcoma (MCS)	Focal destructive infiltration of mast cells	KIT	Solid local tumor consisting of undifferentiated mast cells with high-grade cytology (destructive growth pattern into a terminal phase)
Extracutaneous mastocytoma	Focal nondestructive infiltration of mast cells in sites other than the skin	KIT	Rare benign tumor with low-grade cytology (without destructive growth) primarily in the lungs

[a] SM criteria: Major diagnostic criteria are multifocal, dense aggregates of mast cells (≥15 mast cells aggregating) in bone marrow or other extracutaneous organs; minor diagnostic criteria include (i) >25% mast cells being spindle-shaped or of atypical morphology (type I plus type II); (ii) detection of KIT point mutation at codon D816; (iii) expression of CD25 or CD2 (or both) on mast cells in bone marrow, blood, or other extracutaneous tissues; (iv) persistent serum total tryptase >20 ng/mL. At least one major criterion and one minor criterion or at least three minor criteria must be fulfilled for SM diagnosis.

[b] C-findings: Cytopenia, osteolysis, malabsorption, hypoalbuminemia, weight loss, splenomegaly, hepatomegaly, portal hypertension, and ascites due to widespread mast cell infiltration.

[c] B-findings: Bone marrow mast cell count >30% and/or serum tryptase of >200 ng/mL, bone marrow hypercellularity, dysmyelopoiesis without cytopenia, and organomegaly without functional impairment.

mast cells are mainly found in the skin; whereas in SM, neoplastic mast cells accumulate in multiple organs (e.g., the bone marrow, liver, spleen, lymph nodes, and small intestines) [3–5].

14.3 Epidemiology

Mastocytosis has an annual incidence of 1 case per 100,000 and typically presents as CM in children and SM in adults (an especially indolent variant).

14.4 Pathogenesis

Mastocytosis is linked to mutations in KIT located on a 21-exon containing gene on chromosome 4q12, which encodes a 976 amino acid protein with a molecular weight of 145 kDa. Expressed by hematopoietic stem/progenitor cells, germ cells, melanocytes, and interstitial cells of Cajal, this protein functions as a type III tyrosine kinase transmembrane receptor (CD117). The ligand for KIT receptor, stem cell factor (SCF), or KIT ligand induces and regulates the development, proliferation, maturation, survival, and mediator release from mast cells. Mutations in the KIT gene lead to a gain-of-function receptor and result in uncontrolled proliferation, enhanced survival, accumulation and degranulation of mast cells, and subsequent infiltration of various tissues. KIT gene mutation (especially D816V) is detected in more than 80% of all systemic mastocytosis cases (>90% in indolent and >70% in advanced disease) [3–5].

Apart from KIT mutations, other somatic defects (e.g., TET2, SRSF2, ASXL1, RUNX1, JAK2, and RAS mutations) are observed in advanced systemic mastocytosis cases [4].

A combination of neoplastic mast cell growth and release of mediators (e.g., osteoblast-promoting histamine, heparin, and osteoclast-activating prostaglandin D2) contribute to the effects of mastocytosis in the bone and other organs.

Triggers for mastocytosis include physical factors (heat/cold, sunlight, temperature change, rubbing or pressure on a skin lesion), emotional factors (stress/anxiety, sleep deprivation), drugs (aspirin, morphine, cough medication, alcohol, local anesthetics, β-blockers, anticholinergic medication, vancomycin, amphotericin B, and vitamin B1), venoms (Hymenoptera, snake venom), infectious diseases with fever, and other agents (dental and endoscopic procedures, vaccines, surgery, and contrast media—particularly iodine).

14.5 Clinical features

In cutaneous mastocytosis, clinical symptoms include abnormal growths (bumps, spots, blister lesions) on the skin and erythema and urticaria around the lesion. In systemic mastocytosis, patients may experience episodes (lasting for 15–30 min) of severe symptoms (e.g., itching, flushing, vomiting, diarrhea, muscle and joint pain, headaches, fatigue, nasal congestion, and hypotension). The accumulation and degranulation of mast cells in the bone may lead to osteoporosis (in 18%–31% of indolent disease) [3,6].

14.6 Diagnosis

Diagnosis of mastocytosis is based on clinical manifestations, serum tryptase level, radiographic investigation in affected site, histopathologic examination of tissue, bone marrow biopsy with immunohistochemical stains, flow cytometric evaluation of CD2 and CD25, and molecular analysis of c-kit [6].

CM accounts for 90% of mastocytosis cases and may appear as maculopapular CM (urticaria pigmentosa), diffuse CM, or solitary mastocytoma. Urticaria pigmentosa lesions are usually reddish brown macules or papules. Diffuse CM lesions may manifest as diffuse xanthogranulomas similar to yellow-orange subcutaneous nodules or diffuse red bullae. A solitary mastocytoma has a single or multiple brown nodules, which may display flushing and systemic symptoms (e.g., hypotension after exposure to friction). Histopathologically, CM shows mast cell accumulation along the papillary dermis and reticular dermis and inside the subcutaneous adipose tissue, without bone marrow or other organ infiltration. Serum tryptase is usually not elevated in CM [7,8].

SM is a heterogeneous disease with neoplastic mast cell accumulation in multiple organs (e.g., bone marrow, liver, spleen, lymph nodes, and small intestines). SM may be separated into ISM (representing 80% of SM, with 20% lacking skin involvement), SSM, SM-AHN, ASM, and MCL (<2% of SM) (Table 14.1).

Immunohistochemically, mastocytosis is positive for tryptase, CD117, CD2 (weak in neoplastic mast cells, strong in surrounding reactive lymphocytes), CD25, and CD30 (with preferential positivity in MCL and ASM more than ISM).

Mastocytosis is differentiated from MCAS by the latter's (i) typical clinical symptoms (e.g., hot flushing or pale skin, shortness of breath, itchy hives, irregular heartbeat, headache, chest pain, nausea, vomiting, diarrhea, dizziness, and fainting), (ii) greater than 20% increase in serum total tryptase (detected by CD117 and CD25) above baseline plus 2 ng/mL during or within 4 h after a symptomatic period, and (iii) clinical response to histamine receptor blockers or mast cell-targeting agents (e.g., cromolyn).

14.7 Treatment

Treatment is generally not needed for isolated CM. Instead, patients are advised to avoid excessively hot baths, sudden temperature changes, physical stimulation, hard rubbing of the skin, stress, bee stings, and drugs (e.g., aspirin, codeine, morphine, alcohol, and radiocontrast substances containing iodide—to prevent exacerbations). Corticosteroid and antihistaminic treatment may be administered if necessary [8–10].

Treatment for ISM centers primarily on symptom relief (e.g., epinephrine for acute anaphylaxis, H1 and H2 receptor blockers for pruritus, flushing and gastric hypersecretion, corticosteroids for intestinal malabsorption, and ascites) [8].

Treatment for ASM aims to reduce mast cell load through cytoreductive therapy (e.g., cladribine, interferon-alpha, and hydroxyurea) [8].

14.8 Prognosis

Mastocytosis in pediatric patients may regress at puberty, but mastocytosis in adults does not regress. Although CM and ISM have favorable prognoses, ASM and MCL have poor prognoses. Median survival for ASM is 41 months, that for SM-AHN is 24 months, and that for MCL is 2 months. Poor prognostic factors are late age of onset, weight loss, high lactate dehydrogenase (LDH), high alkaline phosphatase, hypoalbuminemia, cytopenias, bone marrow hypercellularity, organomegaly, organ dysfunction, and atypical morphology of mast cells.

References

1. Swerdlow SH, Campo E, Harris NL, et al. (eds). *WHO classification of tumours of haematopoietic and lymphoid tissues.* International Agency for Research on Cancer; World Health Organization. Lyon, France: IARC, 2008.
2. Arber DA, Orazi A, Hasserjian R, et al. The 2016 revision to the World Health Organization classification of myeloid neoplasms and acute leukemia. *Blood.* 2016;127(20):2391–405.

3. Carter MC, Metcalfe DD, Komarow HD. Mastocytosis. *Immunol Allergy Clin North Am.* 2014;34(1):181–96.
4. Chatterjee A, Ghosh J, Kapur R. Mastocytosis: A mutated KIT receptor induced myeloproliferative disorder. *Oncotarget.* 2015;6(21):18250–64.
5. Ke H, Kazi JU, Zhao H, Sun J. Germline mutations of KIT in gastrointestinal stromal tumor (GIST) and mastocytosis. *Cell Biosci.* 2016;6:55.
6. Lange M, Ługowska-Umer H, Niedoszytko M, et al. Diagnosis of mastocytosis in children and adults in daily clinical practice. *Acta Derm Venereol.* 2016;96(3):292–7.
7. González-de-Olano D, Matito A, Orfao A, Escribano L. Advances in the understanding and clinical management of mastocytosis and clonal mast cell activation syndromes. *F1000Res.* 2016;5:2666.
8. Onnes MC, Tanno LK, Elberink JN. Mast cell clonal disorders: Classification, diagnosis and management. *Curr Treat Options Allergy.* 2016;3(4):453–64.
9. Ustun C, Arock M, Kluin-Nelemans HC, et al. Advanced systemic mastocytosis: From molecular and genetic progress to clinical practice. *Haematologica.* 2016;101(10):1133–43.
10. Tamay, Özçeker D. Current approach to cutaneous mastocytosis in childhood. *Turk Pediatri Ars.* 2016;51(3):123–7.

15
Myeloid/Lymphoid Neoplasms with Eosinophilia and Rearrangement

15.1 Definition

Classified along with myeloproliferative neoplasms (MPN), mastocytosis, myelodisplastic/myeloproliferative neoplasms (MDS/MPN), myelodisplastic syndrome (MDS), acute myeloid leukemia (AML), blastic plasmacytoid dendritic cell neoplasms (BPDCN), acute leukemia of ambiguous lineage (ALAL), B-lymphoblastic leukemia/lymphoma, and T-lymphoblastic leukemia/lymphoma under the myeloid neoplasms and acute leukemia category (see Chapter 13), myeloid/lymphoid neoplasms with eosinophilia and rearrangement are further separated into: (i) myeloid/lymphoid neoplasms with *PDGFRA* rearrangement, (ii) myeloid/lymphoid neoplasms with *PDGFRB* rearrangement, (iii) myeloid/lymphoid neoplasms with *FGFR1* rearrangement, and (iv) provisional entity—myeloid/lymphoid neoplasms with *PCM1-JAK2* rearrangement (Table 15.1) [1,2].

15.2 Biology

Hematopoietic stem cells in the bone marrow include myeloid and lymphoid stem cells. Myeloid stem cells give rise to myeloid lineage cells (i.e., red blood cells [erythrocyte], white blood cells [eosinophils, basophils, neutrophils, mast cells, and monocytes/macrophages but not lymphocytes], and platelets). Lymphoid stem cells differentiate into T- and B-lymphocytes. Neutrophils, eosinophils, and basophils are collectively known as granulocytes, while mast cells and monocytes are also related to granulocytes.

Occurring in the bone marrow and peripheral blood, myeloid/lymphoid neoplasms with eosinophilia and rearrangement of *PDGFRA*, *PDGFRB*, *FGFR1*, or *PCM1-JAK2* are clonal diseases of hematopoietic stem cells, resulting from genetic and epigenetic alterations at *PDGFRA* (platelet-derived growth factor receptor alpha), *PDGFRB* (platelet-derived growth factor receptor beta), *FGFR1* (fibroblast growth factor receptor 1 gene), or *PCM1-JAK2* (human autoantigen pericentriolar material [*PCM1*] gene and

Table 15.1 Subtypes of Myeloid/Lymphoid Neoplasms with Eosinophilia and Rearrangements

Subtype	Cellular Features	Genetic Mutation	Diagnostic Criteria
Myeloid/lymphoid neoplasm with *PDGFRA* rearrangement	Eosinophilia; increased bone marrow mast cells	Cryptic deletion at 4q12; *FIP1L1-PDGFRA* fusion and 66 other combinations	Myeloproliferative neoplasm with prominent eosinophilia and presence of *FIP1L1-PDGFRA* (del[4][q12q12]) fusion; increased tryptase
Myeloid/lymphoid neoplasm with *PDGFRB* rearrangement	Eosinophilia; monocytosis mimicking CMML	*ETV6-PDGFRB* fusion and 25 other combinations	Myeloproliferative neoplasm with prominent eosinophilia and sometimes neutrophilia or monocytosis and presence of *ETV6-PDGFRB* (t[5;12][q33;p13]) or a variant translocation of PDGFRB
Myeloid/lymphoid neoplasm with *FGFR1* rearrangement	Eosinophilia; monocytosis (33%), blasts (50%), hypercellular bone marrow, dysplastic features	Translocations of 8p11.2 involving *FGFR1* and 14 other genes	(i) myeloproliferative neoplasm with prominent eosinophilia and sometimes neutrophilia or monocytosis or (ii) acute myeloid leukemia or precursor T-cell or precursor B-cell lymphoblastic leukemia/lymphoma (usually associated with eosinophilia) and presence of *ZNF198-FGFR1* (t[8;13][p11;q12]) or other FGFR1fusions in myeloid cells and/or lymphoblasts
Myeloid/lymphoid neoplasm with *PCM1-JAK2* rearrangement	Eosinophilia; bone marrow with left-shifted erythroid predominance, lymphoid aggregates, and myelofibrosis	*PCM1-JAK2* fusion	Presence of *PCM1-JAK2* (t[8;9][p22;p24]) and occasionally *ETV6-JAK2* (t[9;12][p24;p13]) and *BCR-JAK2* (t[9;22][p24;q11])

Janus-activated kinase 2 [*JAK2*] fusion). These genetic changes affect the proliferation, differentiation, and self-renewal of eosinophils and lead to the release of granular contents [3,4].

PDGFRA, PDGFRB, FGFR1, or *PCM1-JAK2*-rearranged neoplasms are postulated to arise from a pluripotent stem cell possessing the potential of multilineage differentiation into B, T, or myeloid lineages [5,6].

15.3 Epidemiology

Myeloid/lymphoid neoplasms with eosinophilia and rearrangement of *PDGFRA, PDGFRB, FGFR1,* or *PCM1-JAK2* are rare, with an age-adjusted incidence rate of 0.036 per 100,000. Neoplasms with *FIP1L1-PDGFRA* rearrangement or a myeloproliferative variant of hypereosinophilia (HES) have a peak age of 65–74 years and a male to female ratio of 1.47:1. Neoplasms with *FGFR1* rearrangement (with 100 cases reported) have a median age of 44 years (range 3–84) and a male to female ratio of 1.2:1. Neoplasms with *PCM1-JAK2* rearrangement have a median age of 47 years (range 12–75) and a male to female ratio of 5:1.

15.4 Pathogenesis

A myeloid/lymphoid neoplasm with eosinophilia and rearrangement of *PDGFRA* typically occurs as *FIP1L1-PDGFRA* fusion, which is generated from a submicroscopic 800-kb interstitial deletion on chromosome 4, del(4)(q12q12) (Table 15.1). The fusion disrupts the autoinhibitory juxtamembrane domain of *PDGFRA*, resulting in constitutive activation of disparate tyrosine kinase (TK) gene [4,5].

A myeloid/lymphoid neoplasm with eosinophilia and rearrangement of *PDGFRB* often harbors *ETV6-PDGFRB* [t(5;12)(q33;p13)], *H4-PDGFRB, CCDC88C-PDGFRB,* or other fusion combinations. Because it is a class III receptor tyrosine kinase located at chromosomal position 5q31-q33, a fusion translocation involving *PDGFRB* typically contributes to protein dimerization and consequently to the constitutive activation of the *PDGFRB* kinase domain (Table 15.1) [4,5].

A myeloid/lymphoid neoplasm with eosinophilia and rearrangement of *FGFR1* contains rearrangement of *FGFR1* on chromosome 8p11 to at least 14 partner genes, such as *ZNF198-FGFR1* (t[8;13][p11;q12], 40%), *BCR-FGFR1* (t[8;22][p11;q11],18%), *CEP110-FGFR1* (t[8;9][p11;q33], 15%), and *FGFR1OP1-FGFR1* (t[6;8][q27;p11–12], 9%). *FGFR1* belongs to a family of four high-affinity receptor tyrosine kinases (*FGFR1–4*) that dimerize upon ligand binding, activating multiple signaling pathways. *FGFR1* aberration contributes to cell transformation [4,6].

A myeloid/lymphoid neoplasm with eosinophilia and rearrangement of *PCM1-JAK2* and occasionally *ETV6-JAK2* (t[9;12][p24;p13]) and *BCR-JAK2* (t[9;22][p24;q11]) usually presents with features of a chronic myeloid neoplasm such as eosinophilia and/or bone marrow fibrosis. It may undergo rapid progression from chronic phase disease to AML and rarely to lymphoid blast phase [4,7].

15.5 Clinical features

Clinical presentation of myeloproliferative neoplasms ranges from eosinophilia and lymphadenopathy to T-cell non-Hodgkin lymphoma (that may progress to AML). The most common symptoms are weakness and fatigue (26%), cough (24%), dyspnea (16%), myalgia or angioedema (14%), eosinophilia (12%), rash or fever (12%), rhinitis (10%), night sweats, and fever. Some patients may be asymptomatic.

15.6 Diagnosis

Diagnosis of eosinophilia-associated neoplasms involves blood count/morphology (for blasts, monocytosis, basophilia, left-shifted leukocytosis, dysplasia, or leukoerythroblastosis), serum chemistry (for tryptase or vitamin B12), bone marrow morphology/immunohistochemistry (e.g., CD117, tryptase, and CD25), flow cytometry of myeloid, B- and/or T-lymphocyte markers, and molecular analysis of gene fusions (e.g., real-time PCR or interphase/metaphase FISH).

A myeloid/lymphoid neoplasm with *PDGFRA* rearrangement most commonly appears as chronic eosinophilic leukemia (CEL), AML, T lymphoblastic lymphoma, and CML as well as idiopathic hypereosinophilia (10%–20%). Peripheral eosinophilia >1 × 10^3/μL lasts for more than 6 months.

A myeloid/lymphoid neoplasm with *PDGFRB* rearrangement most commonly manifests as chronic myelomonocytic leukemia (CMML) with eosinophilia, as well as CEL, CML, atypical chronic myeloid leukemia (aCML), primary myelofibrosis (PMF), juvenile myelomonocytic leukemia with eosinophilia, acute myeloid leukemia (without eosinophilia), and even chronic basophilic leukemia (CBL).

A myeloid/lymphoid neoplasm with *FGFR1* rearrangement (also known as 8p11 myeloproliferative syndrome, EMS, or stem cell leukemia/lymphoma, SCLL) often co-occurs with AML, T- or B-lymphoblastic leukemia/lymphoma (B-ALL), or a mixed phenotype acute leukemia.

A myeloid/lymphoid neoplasm with *PCM1-JAK2* rearrangement may show hepatosplenomegaly, primary myelofibrosis, CEL, aCML, and bone marrow fibrosis.

Differential diagnoses for myeloid/lymphoid neoplasms with eosinophilia and rearrangement of *PDGFRA*, *PDGFRB*, *FGFR1*, or *PCM1-JAK2* include other eosinophilia-associated identities such as chronic eosinophilic leukemia not otherwise specified (CEL-NOS), HES, and lymphocyte-variant hypereosinophilia.

CEL-NOS is defined by the absence of the *BCR–ABL1* fusion gene and rearrangements involving *PDGFRA/B* and *FGFR1*, the presence of nonspecific cytogenetic/molecular abnormalities, and/or increased myeloblasts, along with the exclusion of other neoplasms associated with eosinophilia (e.g., AML, B-ALL with t[5;14], SM, CML, PV, ET, and PMF; see Chapters 13 and 14).

HES is defined by persistent eosinophilia ($\geq 1.5 \times 10^9$/L) for more than 6 months and tissue damage. It is considered a provisional diagnosis until a primary or secondary cause of eosinophilia (e.g., reactive eosinophilia, lymphocyte-variant hypereosinophilia, CEL-NOS, eosinophilia-associated MPN, or AML/ALL with rearrangements of *PDGFRA*, *PDGFRB*, or *FGR1*) is identified. If there is no tissue damage, idiopathic HES is the preferred diagnosis.

Lymphocyte variant hypereosinophilia is associated with a cytokine-producing aberrant T-cell population.

15.7 Treatment

Myeloid/lymphoid neoplasms with eosinophilia and rearrangement of *PDGFRA* or *PDGFRB* show excellent response rates to imatinib mesylate and imatinib treatment (100 and 400 mg daily, respectively) for complete and durable hematologic and complete molecular remissions [5,8].

In contrast, neoplasms with *FGFR1* and *JAK2* fusion genes are resistant to imatinib and other clinically available TK inhibitors (TKI). Use of midostaurin or allogeneic hematopoietic stem cell transplantation (HSCT) is necessary for long-term disease-free survival [5,8].

HES (excluding all other possible causes) may be treated with corticosteroids (e.g., prednisone 1 mg/kg) for eosinophil reduction and hydroxyurea for controlling leukocytosis and eosinophilia in steroid nonresponders.

IFN-α helps reverse organ injury in patients with HES and CEL. The second- and third-line agents include vincristine, cyclophosphamide, etoposide, 2-chlorodeoxyadenosine, cytarabine, and cyclosporin-A.

15.8 Prognosis

Neoplasms with *PDGFRA or PDGFRB* rearrangement show high response rates to imatinib treatment and have a 10-year overall survival rate of 90%. Neoplasms with rearranged *FGFR1* follow an aggressive course with a median survival time of less than 12 months and usually terminate in AML in 1–2 years. *PCM1-JAK2* fusion is associated with a poor prognosis, and early allogeneic HSCT should be considered for eligible patients. HES has a 5-year survival rate of 80%, and a 15-year survival rate of 42%.

References

1. Swerdlow SH, Campo E, Harris NL, et al. (eds). *WHO classification of tumours of haematopoietic and lymphoid tissues.* International Agency for Research on Cancer; World Health Organization. Lyon, France: IARC, 2008.
2. Arber DA, Orazi A, Hasserjian R, et al. The 2016 revision to the World Health Organization classification of myeloid neoplasms and acute leukemia. *Blood.* 2016;127(20):2391–405.
3. Reiter A, Walz C, Watmore A, et al. The t(8;9)(p22;p24) is a recurrent abnormality in chronic and acute leukemia that fuses PCM1 to JAK2. *Canc Res.* 2005;65(7):2662–7.
4. Savage N, George TI, Gotlib J. Myeloid neoplasms associated with eosinophilia and rearrangement of *PDGFRA*, *PDGFRB*, and *FGFR1*: A review. *Int J Lab Hematol.* 2013;35(5):491–500.
5. Gotlib J. World Health Organization-defined eosinophilic disorders: 2015 update on diagnosis, risk stratification, and management. *Am J Hematol.* 2015;90(11):1077–89.
6. Andrei M, Bandarchuk A, Abdelmalek C, Kundra A, Gotlieb V, Wang JC. PDGFR$^{\beta}$-rearranged myeloid neoplasm with marked eosinophilia in a 37-year-old man; and a literature review. *Am J Case Rep.* 2017;18:173–80.
7. Troadec E, Dobbelstein S, Bertrand P, et al. A novel t(3;13)(q13;q12) translocation fusing FLT3 with GOLGB1: Toward myeloid/lymphoid neoplasms with eosinophilia and rearrangement of FLT3? *Leukemia.* 2017;31(2):514–17.
8. Reiter A, Gotlib J. Myeloid neoplasms with eosinophilia. *Blood.* 2017;129(6):704–14.

16 Myelodisplastic/Myeloproliferative Neoplasms (MDS/MPN)

16.1 Definition

Classified along with myeloproliferative neoplasms (MPN), mastocytosis, myeloid/lymphoid neoplasms with eosinophilia and rearrangements, myelodisplastic syndrome (MDS), acute myeloid leukemia (AML), blastic plasmacytoid dendritic cell neoplasms (BPDCN), acute leukemia of ambiguous lineage (ALAL), B-lymphoblastic leukemia/lymphoma, and T-lymphoblastic leukemia/lymphoma under the myeloid neoplasms and acute leukemia category (see Chapter 13), myelodisplastic/myeloproliferative neoplasms (MDS/MPN) are an overlapping group of myeloid disorders that include chronic myelomonocytic leukemia (CMML), atypical chronic myeloid leukemia (aCML)—*BCR-ABL1–*, juvenile myelomonocytic leukemia (JMML), MDS/MPN with ring sideroblasts and thrombocytosis (MDS/MPN-RS-T), and MDS/MPN unclassifiable (MDS/MPN-UC) (Table 16.1) [1,2].

16.2 Biology

MDS/MPN are clonal myeloid disorders that possibly evolve from a pluripotent lymphoid-myeloid stem cell or a more committed myeloid progenitor. CMML appears to arise within the CD34(+)/CD38(–) cells and undergoes subsequent granulo-monocytic differentiation. As MDS/MPN display both dysplastic and proliferative features, they cannot be classified conclusively as either myelodysplastic syndromes (MDS) or chronic myeloproliferative disorders (CMPD).

CMML, JMML, and aCML represent three major myeloid disorders within the MDS/MPN group. MDS/MPN-RS-T is considered a provisional identity. MDS/MPN-UC is a "by exclusion" identity that does not meet the criteria for any of the three major MDS/MPN disorders.

Table 16.1 Characteristics of Myelodisplastic/Myeloproliferative Neoplasm (MDS/MPN) Subtypes

Subtype	Cellular Features	Diagnostic Criteria
Chronic myelomono-cytic leukemia (CMML)—*BCR-ABL1–*	Blasts, neutrophil precursors; hypercellularity (75%); granulocytic proliferation; monocytic proliferation; dyserythropoiesis; micromegakaryocytes/megakaryocytes (80%), fibrosis (30%)	(i) Persistent monocytosis is >1 × 10^9/L in PB; no Philadelphia chromosome or *BCR/ABL* fusion gene; (ii) <20% blasts in PB or BM; (iii) dysplasia involving one or more myeloid lineages or, either an acquired clonal cytogenetic BM abnormality or at least 3 months of persistent PB monocytosis (monocytes >10% WBC), if all other causes are ruled out
Atypical chronic myeloid leukemia (aCML)	PB: blasts, immature neutrophils; monocytes; thrombocytopenia. BM: granulocytic hypercellularity; blasts; dysgranulopoiesis; megakaryocytic dysplasia; erythroid precursors	PB leukocytosis; prominent dysgranulopoiesis; no Philadelphia chromosome or *BCR/ABL* fusion; neutrophil precursors (e.g., promyelocytes, myelocytes, and metamyelocytes) >10%, basophils <2%, monocytes <10% of WBC; granulocytic hypercellularity and dysplasia in BM; <20% blasts; thrombocytopenia (<100 × 10^9/L)
Juvenile myelomonocytic leukemia (JMML)	PB: leukocytosis (neutrophils, promyelocytes, myelocytes, and monocytes); blasts; thrombocytopenia. BM: hypercellularity with granulocytes and erythroid precursors; monocytes (5%–10%)	Major criteria: (i) no Philadelphia chromosome or *BCR/ABL* fusion gene; (ii) PB monocytosis >1 × 10^9/L; (iii) <20% blasts in PB and BM. Minor criteria: (i) fetal hemoglobin (Hb F) increase; (ii) immature granulocytes in PB; (iii) WBC >1 × 10^9/L; (iv) clonal chromosomal abnormality (e.g., monosomy 7); (v) granulocyte-macrophage colony-stimulating factor (GM-CSF) hypersensitivity of myeloid progenitors in vitro
MDS/MPN with ring sideroblasts and thrombocytosis (MDS/MPN-RS-T)	Ring sideroblasts, dyserythropoiesis, thrombocytosis, megakaryocytes	Persistent BM ring sideroblasts (>15%) and sustained thrombocytosis (platelet count >450 × 10^8/dL)

(Continued)

Table 16.1 *(Continued)* Characteristics of Myelodisplastic/Myeloproliferative Neoplasm (MDS/MPN) Subtypes

Subtype	Cellular Features	Diagnostic Criteria
MDS/MPN unclassifiable (MDS/MPN-UC)	PB: dimorphic erythrocytes; thrombocytosis; leukocytosis. BM: basophilia, megakaryocytic hyperplasia and intense fibrosis	(i) features of myelodysplastic syndrome (MDS) with <20% blasts; (ii) prominent myeloproliferative features; (iii) CMPD, MDS, or recent cytotoxic or growth factor therapy; (iv) no Philadelphia chromosome or *BCR/ABL* fusion gene, del(5q), t(3;3)(q21;q26), or inv(3)(q21q26); (v) inability to assign to other MDS, CMPD, or MDS/MPN

Note: BM, bone marrow; PB, peripheral blood.

16.3 Epidemiology

MDS/MPN have an estimated incidence of up to 3 cases per 100,000. Specifically, CMML has an annual incidence of 1 per 100,000, a median age at diagnosis of 70 years, and a male predominance (male to female ratio of 1.5–3.1:1); aCML is an extremely rare subtype with a median age at diagnosis in the seventh or eighth decade of life. JMML (also known as juvenile chronic myelomonocytic leukemia) has an incidence of 0.12 per 100,000 children and accounts for 2% of all childhood leukemia. It typically affects young children (median age, 2 years) and shows a male to female ratio of 2.5:1. MDS/MPN-U constitutes <5% of all myeloid malignancies and has a median age of 71 years and a male predominance (male to female ratio of 2:1).

16.4 Pathogenesis

Risk factors for CMML include exposure to occupational and environmental carcinogens, ionizing radiation, and cytotoxic agents; those for JMML are neurofibromatosis type 1 (NF1, up to 14% cases) and Noonan syndrome (RAS signaling abnormalities).

Molecularly, CMML contains somatic mutations of signal transduction (*NRAS, KRAS, CBL, and JAK2*), DNA methylation (*DNMT3A, TET2, and IDH 1/2*), transcriptional regulation (*ETV6 and RUNX1*), chromatin modification (*EZH2 and ASXL1*), and the RNA splicing machinery (*SF3B1, SRSF2, ZRSR2, and U2AF1*) in addition to cytogenetic abnormalities of trisomy 8, monosomy 7, del(7q), and rearrangements with a 12p breakpoint.

Sequencing analysis of nine genes with recurrent mutations (e.g., *TET2* [50–60%], *SRSF2* [40–50%], *ASXL1* [35–40%]) enables the determination of a genetic clonal event in >90% of CMML cases and differentiation of malignancy from reactive monocytosis [3–5].

aCML is linked to mutations in *SETBP1* (chromosome 18q21, 25%), *JAK2, NRAS, IDH2, CBL, CSF3R*, and *ETNK1*, as well as to cytogenetic abnormality of i(17q). Analysis of these mutated genes permits identification of a clonal event in >50% of aCML cases [3–5].

JMML harbors mutations in the RAS pathway (90%, *PTPN11* [Noonan syndrome] more than *NF1* [neurofibromatosis type 1] more than *NRAS/KRAS* more than *CBL*) as well as secondary events (*SETBP1* and *JAK3*). Additionally, JMML demonstrates characteristic granulocyte-macrophage-colony-stimulating factor (GM-CSF) hypersensitivity [3–5].

MDS/MPN-RS-T has mutations in *SF3B1* (72%), *JAK2*V617F (50%), *MPL, TET2, ASXL1*, and *EZH2* [3–5].

MDS/MPN-U also includes *JAK2*V617F-positivity (20%–30%) and genetic abnormalities of aneuploidies (trisomy 8 and monosomy 7) or deletions (del7q, del13q, and del20q) in addition to reciprocal translocations involving diverse tyrosine kinase (TK) fusion genes [3–5].

16.5 Clinical features

CMML may present clinically with (i) fever, fatigue, night sweats, and weight loss; (ii) infection (pyoderma gangrenosum and vasculitis); (iii) bleeding (due to idiopathic thrombocytopenia); (iv) hepatomegaly; (v) splenomegaly; (vi) skin and lymph node infiltration; and (vii) serous membrane effusions.

aCML is associated with (i) anemia; (ii) thrombocytopenia; and (iii) splenomegaly (75%).

JMML may display (i) constitutional symptoms (e.g., malaise, pallor, and fever), (ii) bronchitis or tonsillitis (50%), (iii) bleeding diathesis, (iv) maculopapular skin rashes (40%–50%), (v) lymphadenopathy (75%), and (vi) hepatosplenomegaly.

MDS/MPN-RS-T may cause anemia and moderate splenomegaly.

MDS/MPN-UC may show (i) features of both MDS and CMPD, (ii) hepatomegaly, and (iii) splenomegaly.

16.6 Diagnosis

The diagnostic criteria for various MDS/MPN subtypes are summarized in Table 16.1.

As the most frequent subtype of MDS/MPN, CMML is defined by persistent peripheral monocytosis (>1 × 10^9/L, with monocytes containing bizarre nuclei and cytoplasmic granules and accounting for >10% white blood cells [WBCs]) and dysplasia.

aCML is characterized by left-shifted leukocytosis with severe neutrophil dysplasia and cytopenias and an absence of monocytosis or the *BCR-ABL1* fusion typical of CML.

JMML is characterized by an overproduction of monocytes that infiltrate the liver, spleen, lung, intestine, and other organs, which may lead to considerable morbidity and mortality.

MDS/MPN-RS-T is defined by persistent ringed sideroblasts and sustained thrombocytosis.

MDS/MPN-U is defined by a combination of dysplastic and myeloproliferative features that do not fulfill criteria for assignment to any other MDS/MPN subtype.

16.7 Treatment

CMML may be treated with topotecan (topoisomerase I inhibitor), cytarabine (pyrimidine-analog antimetabolite), hydroxyurea, and nucleoside 5-azacitidine (inhibitor of DNA methyltransferase). Imatinib mesylate is effective in a subset of CMML related to *PDGFRB* fusion oncogenes. Hypomethylating agents (HMAs) such as azacitidine may be used in CMML with cytopenias as the predominant symptom. Allogeneic hematopoietic stem cell transplantation (allo-HSCT, which requires a match of greater than or equal to three loci with the HLA gene) or bone marrow transplantation (BMT) is the only treatment modality providing long-term remission and a potential cure [6–8].

aCML benefits from the use of HMAs, hydroxyurea, and lenalidomide. Allo-HSCT appears to be the treatment offering a curative outcome.

JMML with CBL mutations and its associated congenital syndrome may be self-limited, while RAS-mutated JMML requires allogeneic transplant using sibling or unrelated human leukocyte antigen (HLA)-matched donor marrow or autologous marrow. JMML responds poorly to chemotherapy, and >90% of patients die despite such treatment.

MDS/MPN-RS-T treatment aims to address anemia (erythropoiesis stimulating agents and/or transfusion) and thrombotic risk (aspirin).

MDS/MPN-UC can be treated with imatinib mesylate at a standard dosage. Other potentially useful drugs consist of HMAs, interferon alpha, cyclosporine, thalidomide, lenalidomide, and anti-thymocyte globulin.

16.8 Prognosis

CMML has a median survival of 12–24 months, and a 3-year overall survival rate of 32%. The incidence of CMML transformation into AML is 15%–52% (especially those with mutation in *ASXL1* or *RUNX1*). Unfavorable prognostic factors include low hemoglobin level and platelet counts; high WBC, monocyte, and lymphocyte counts; the presence of circulating immature myeloid cells; a high percentage of marrow blasts; a low percentage of marrow erythroid cells; abnormal cytogenetics; and high levels of serum LDH and beta-2-microglobulin.

aCML has a median survival of 12.4 months and may evolve to acute leukemia (25%–40%). Fatal complications include resistant leukocytosis, anemia, thrombocytopenia, hepatosplenomegaly, cerebral bleeding (due to thrombocytopenia), and infection.

JMML has a 5-year survival rate of 52% and may evolve to acute leukemia (10%–20%). Prognosis is better for children under 1 year than children more than 2 years of age at the time of diagnosis. Low platelet counts and high Hb F levels are associated with a worse prognosis.

MDS/MPN-RS-T has a median survival of 3.3 years compared to 6.9 years in *SF3B1* mutated cases. MDS/MPN-U has a median survival rate of 21.8 months.

References

1. Swerdlow SH, Campo E, Harris NL, et al. (eds). *WHO classification of tumours of haematopoietic and lymphoid tissues*. International Agency for Research on Cancer; World Health Organization. Lyon, France: IARC, 2008.

2. Arber DA, Orazi A, Hasserjian R, et al. The 2016 revision to the World Health Organization classification of myeloid neoplasms and acute leukemia. *Blood.* 2016;127(20):2391–405.
3. Li B, Gale RP, Xiao Z. Molecular genetics of chronic neutrophilic leukemia, chronic myelomonocytic leukemia and atypical chronic myeloid leukemia. *J Hematol Oncol.* 2014;7:93.
4. Odenike O, Onida F, Padron E. Myelodysplastic syndromes and myelodysplastic/myeloproliferative neoplasms: An update on risk stratification, molecular genetics, and therapeutic approaches including allogeneic hematopoietic stem cell transplantation. *Am Soc Clin Oncol Educ Book.* 2015:e398–412.
5. Mughal TI, Cross NC, Padron E, et al. An International MDS/MPN Working Group's perspective and recommendations on molecular pathogenesis, diagnosis and clinical characterization of myelodysplastic/myeloproliferative neoplasms. *Haematologica.* 2015;100(9):1117–30.
6. Savona MR, Malcovati L, Komrokji R, et al. An international consortium proposal of uniform response criteria for myelodysplastic/myeloproliferative neoplasms (MDS/MPN) in adults. *Blood.* 2015;125(12):1857–65.
7. Clara JA, Sallman DA, Padron E. Clinical management of myelodysplastic syndrome/myeloproliferative neoplasm overlap syndromes. *Canc Biol Med.* 2016;13(3):360–72.
8. PDQ Adult Treatment Editorial Board. *Myelodysplastic/ Myeloproliferative Neoplasms Treatment (PDQ®): Health Professional Version.* PDQ Cancer Information Summaries [Internet]. Bethesda, MD: National Cancer Institute (US); 2002–2017.

17
Myelodysplastic Syndromes (MDS)

17.1 Definition

Myelodysplastic syndromes (MDS) fall under the myeloid neoplasms and acute leukemia category, along with myeloproliferative neoplasms (MPN), mastocytosis, myeloid/lymphoid neoplasms with eosinophilia and rearrangements, myelodysplastic/myeloproliferative neoplasms (MDS/MPN), acute myeloid leukemia (AML), blastic plasmacytoid dendritic cell neoplasms (BPDCN), acute leukemia of ambiguous lineage (ALAL), B-lymphoblastic leukemia/lymphoma, and T- lymphoblastic leukemia/lymphoma (see Chapter 13) [1,2].

As a group of heterogeneous clonal hematopoietic stem cell disorders characterized by ineffective hematopoiesis, progressive cytopenia, and transformation to acute myeloid leukemia (AML) in high-risk cases, MDS are further divided into (i) MDS with dysplasia (including single lineage dysplasia MDS-SLD and multilineage dysplasia MDS-MLD), (ii) MDS with ring sideroblasts (including single lineage dysplasia MDS-RS-SLD and multilineage dysplasia MDS-RS-MLD), (iii) MDS with excess blasts (MDS-EB, including subtypes 1 and 2), (iv) MDS with isolated del(5q), (v) MDS unclassifiable (MDS-U, including 1% blood blasts and single lineage dysplasia/pancytopenia), (vi) provisional entity—refractory cytopenia of childhood, and (vii) myeloid neoplasms with germline predisposition [2]. This morphological classification is largely based on the percentage of myeloblasts in the bone marrow and peripheral blood, the type and degree of myeloid dysplasia, and the presence of ring sideroblasts (Table 17.1) [2,3].

Among MDS subtypes, MDS-EB subtype 2 accounts for 40% of MDS cases, MDS-MLD for 30%, MDS-SLD for 10%–20%, and MDS-RS for 3%–11%.

17.2 Biology

Being a soft, spongy, gelatinous tissue within the hollow spaces of the bones, bone marrow is separated into red marrow and yellow marrow.

Table 17.1 Cellular and Cytogenetic Characteristics of Myelodysplastic Syndromes (MDS)

MDS Subtype	Peripheral Blood	Bone Marrow	Cytogenetics by Conventional Karyotype Analysis
MDS with single lineage dysplasia (MDS-SLD)	Cytopenia, blasts <1%	Blasts <5%, unilineage dysplasia in ≥10% cells; ringed sideroblasts <15% of erythroid precursors	Any, unless MDS with isolated del(5q)
MDS with multilineage dysplasia (MDS-MLD)	Cytopenia, blasts <1%, no Auer rods; monocytes <1 × 10^6/μL	Blasts <5%, dysplasia ≥10% cells (at least 2 myeloid lineages), no Auer rods, ringed sideroblasts <15% of erythroid precursors	Any, unless MDS with isolated del(5q)
MDS-RS with single lineage dysplasia (MDS-RS-SLD)	Blasts <1%	Blasts <5%, ringed sideroblasts ≥15% of erythroid precursors	Any, unless MDS with isolated del(5q)
MDS-RS with multilineage dysplasia (MDS-RS-MLD)	Blasts <1%	Blasts <5%, ringed sideroblasts ≥15% of erythroid precursors	Any, unless MDS with isolated del(5q)
MDS with excess blasts (MDS-EB) subtype 1	Cytopenia, blasts 2-4%, no Auer rods, monocytes <1 × 10^6/μL	Blasts 5-9%, dysplasia, no Auer rods	Any
MDS with excess blasts (MDS-EB) subtype 2	Cytopenia, blasts 5-19%, no Auer rods, monocytes <1 × 10^6/μL	Blasts 10-19%, dysplasia, Auer rods	Any
MDS with isolated del(5q)	Anemia, normal to elevated platelet count, blasts <1%	Blasts <5%, normal to elevated megakaryocytes with hypolobated nuclei, no Auer rods	del(5q) alone or with 1 additional abnormality except −7 or del(7q)
MDS-unclassified (MDS-U)	Cytopenia, blasts ≤1%	Blasts <5%, dysplasia in <10% of at least one myeloid cell lines	Any

(Continued)

Table 17.1 *(Continued)* Cellular and Cytogenetic Characteristics of Myelodysplastic Syndromes (MDS)

MDS Subtype	Peripheral Blood	Bone Marrow	Cytogenetics by Conventional Karyotype Analysis
Provisional identity - refractory cytopenia of childhood	Blasts <2%	Blasts <5%	Any
Myeloid neoplasms with germline predisposition	Blasts <1%	Blasts <5%, ringed sideroblasts <15% of erythroid precursors	MDS-defining abnormality

Note: Cytopenia is defined as having hemoglobin <10 g/dL, platelet count <100 × 10^9/L, and absolute neutrophil count <1.8 × 10^9/L.

Red marrow is a hematopoietic tissue in which hematopoietic stem cells generate various blood cells (e.g., erythrocytes, granulocytes, monocytes, thrombocytes, and lymphocytes). Yellow marrow is a fatty/fibrous tissue (also called stroma) in which mesenchymal stem cells generate osteoblasts, osteoclasts, chondrocytes, myocytes, fibroblasts, macrophages, adipocytes, and endothelial cells, providing the microenvironment and colony-stimulating factors for hematopoiesis by hematopoietic tissue.

Red marrow predominates in the flat bones, including the hip bone, sternum/breast bone, skull, ribs, vertebrae, shoulder blades, and the metaphyseal and epiphyseal ends of the long bones (e.g., the femur, tibia, and humerus), where the bone is cancellous or spongy. Yellow marrow mainly occurs in the hollow interior of the diaphyseal portion or the shaft of long bones and often replaces red marrow in old age. However, yellow marrow may revert to red marrow in the event of increased demand for red blood cells (e.g., blood loss).

MDS results largely from disturbances of differentiation and maturation and from changes in bone marrow stroma that lead to excessive apoptosis of hematopoietic precursors, ineffective bone marrow hematopoiesis, peripheral blood cytopenia, and subsequent progression to AML (30%).

Approximately 90% of MDS cases arise de novo without identifiable cause, and complex epigenetic, genetic, and immunologic mechanisms appear to underline MDS pathogenesis.

17.3 Epidemiology

MDS is a relatively common hematological malignancy, with an annual incidence of 5 per 100,000 in the general population, 30 per 100,000 in the age group of greater than 60 years, and 50 per 100,000 in the age group of greater than 80 years. The median age of diagnosis is 65 to 70 years, and <10% of patients are under 50 years. There is a male predominance among MDS patients (1.5:1).

17.4 Pathogenesis

Risk factors for MDS include advancing age; exposure to tobacco smoke, ionizing radiation, organic chemicals (e.g., benzene, toluene, xylene, and chloramphenicol), heavy metals, herbicides, pesticides, fertilizers, stone and cereal dusts, exhaust gases, nitro-organic explosives, and petroleum and diesel derivatives; chemotherapy or radiation therapy for another malignancy (10%–15% of MDS cases); and genetic syndromes (e.g., Fanconi's anemia, dyskeratosis congenita, Down syndrome, and familial platelet disorders) [4,5].

Molecularly, MDS is linked to cytogenetic abnormalities (e.g., deletions of the long arm of chromosome 5 [del5q, affecting the q31 to q33 bands], monosomy Y, monosomy 7 [del7] or its long arm [del7q], trisomy 8, and del20q); genetic alterations involving DNA methylation (*TET2* [on 4q24, 21% cases] *DNMT3A*, *IDH2* [on 15q26, 2% cases], *IDH1* [on 2Q33, 1% cases]), post-translational chromatin modification (*ASXL1* [on 20q11, 14% cases], *EZH2* [on 7q36, 6% cases]), RNA spliceosome machinery (*SF3B1* [on 2q33, 28% cases], *SRSF2* [on 17q25, 12% cases], *ZRSR2*, *U2AF1* [on 21q22, 7% cases]), transcription regulation (*RUNX1* [on 21q22, 9% cases], *TP53* [on 17q13, 8% cases], *ETV6* [on 12p13, 3% cases], *BCOR*, *PHF6*, *NCOR*, *CEBPA*, *GATA2*), tyrosine kinase receptor signaling (*JAK2* [on 9p24, 3% cases], *MPL*, *FLT3*, *GNAS*, *KIT*]), RAS pathways (*KRAS*, *NRAS* [on lp13, 4% cases], *CBL* [on 11q23, 2% cases], *NF1*, *PTPN11*), DNA repair (*ATM*, *BRCC3*, *DLRE1C*, *FANCL*), and cohesion complexes (*STAG2*, *CTCF*, *SMC1A*, *RAD21*); and immunologic aberrations (e.g., aberrant immune attack on myeloid progenitors resulting in increased apoptosis) [4–7].

17.5 Clinical features

Clinical manifestations of MDS range from anemia (85% of cases); bleeding, easy bruising, fatigue, splenomegaly (10%); to hepatomegaly (5%).

17.6 Diagnosis

Diagnosis of MDS centers on the cytological examination of peripheral blood smears, evaluation of bone marrow aspirates and bone marrow biopsy (for cellularity and fibrosis), analysis of conventional cytogenetics, immunophenotyping/flow cytometry, and FISH/molecular investigations.

Patients (especially the elderly) having an abnormal blood count (e.g., macrocytic anemia with significantly reduced reticulocyte, neutropenia, and thrombocytopenia) are highly suspected of MDS and should undergo bone marrow evaluation to exclude all causes of reactive cytopenia/dysplasia.

The current diagnostic criteria for MDS consist of (i) persistent (more than 6 months' duration) and significant cytopenia (i.e., hemoglobin<10 g/dL, absolute neutrophil count<1.8×10^9/L, platelet count<100×10^9/L); (ii) significant bone marrow dysplasia (at least 10% of major hematopoietic lineages), blast excess (at least 15% ring sideroblasts or 5%–19% myeloblasts in bone marrow smears), or typical cytogenetic abnormality; and (iii) exclusion of differential diagnoses (Table 17.1).

It should be noted that MDS with isolated del(5q) is defined by an isolated del(5q) in the absence of excess blasts; MDS with single lineage dysplasia and pancytopenia or 1% peripheral blasts on at least two occasions is regarded as MDS-U; and the presence of >20% blasts suggests acute myeloid leukemia with myelodysplasia-related changes (AML-MRC).

In addition, because cytopenia, dysplasia, bone marrow ring sideroblasts, increased myeloblasts, and chromosomal abnormalities are not specific for MDS, a thorough history and physical examination are required to rule out conditions that cause pancytopenia and mimic MDS, particularly vitamin B12/folate and zinc/copper deficiency. Other differential diagnoses include autoimmune cytopenia, AML, myeloproliferative neoplasms, aplastic anemia, paroxysmal nocturnal hemoglobinuria, and large granular lymphocytic (LGL) leukemia.

17.7 Treatment

Treatment options for MDS consist of hypomethylating agents (HMA, including azacitidine and decitabine for all subgroups of MDS), lenalidomide (an inhibitor of CSK1A1 serine/threonine kinase 5q32 for MDS with 5q deletion and lower-risk disease that requires red cell transfusions), and allogeneic hematopoietic stem cell transplantation (HSCT). HMA show a response rate of about 40% in high-risk patients and a median response duration of 9–15 months. Lenalidomide has response rates of 47% in MDS without del(5q) and 83% in MDS with del(5q), resulting in cytogenetic improvement and achieving transfusion independence (70% cases) or complete cytogenetic remission in 4.6 weeks. Allogeneic HSCT is potentially curative for 25%–60% of patients but is restricted by age and donor availability [8].

Because anemia occurs in more than 90% of MDS cases at diagnosis or during the course of the disease, transfusion support is often necessary. Erythropoiesis stimulating agents (ESAs, e.g., erythropoietin [EPO] and darbepoetin) are indicated for patients with hemoglobin <10 g/dL and an erythropoietin level <500 mU/mL. The anabolic steroid danazol may be also considered for anemic patients [8].

17.8 Prognosis

The International Prognostic Scoring System (IPSS) stratifies MDS patients into four risk groups: low (33% cases), intermediate-1 (38% cases), intermediate-2 (22% cases), and high risk (7% cases), which have median overall survival times of 5.7, 3.5, 1.1, and 0.4 years, respectively, in the absence of therapy. Furthermore, the median time for 25% of patients with low, intermediate-1, intermediate-2, and high-risk MDS to progress to AML is 9.4, 3.3, 1.1, and 0.2 years, respectively, in the absence of therapy [5].

Patients with HMA nonresponding MDS have a median survival of less than 6 months. Patients with defects in TET2 respond well to azacitidine, while those with IDH and DNMT3A mutations have a good response to decitabine. Patients with a del5q as well as TP53 mutation respond poorly to lenalidomide. AML is notably refractory to chemotherapy.

References

1. Swerdlow SH, Campo E, Harris NL, et al. (eds). *WHO classification of tumours of haematopoietic and lymphoid tissues.* International Agency for Research on Cancer; World Health Organization. Lyon, France: IARC, 2008;1–439.
2. Arber DA, Orazi A, Hasserjian R, et al. The 2016 revision to the World Health Organization classification of myeloid neoplasms and acute leukemia. *Blood.* 2016;127(20):2391–405.
3. Gangat N, Patnaik MM, Tefferi A. Myelodysplastic syndromes: Contemporary review and how we treat. *Am J Hematol.* 2016;91(1):76–89.

4. Shahrabi S, Khosravi A, Shahjahani M, Rahim F, Saki N. Genetics and epigenetics of myelodysplastic syndromes and response to drug therapy: New Insights. *Oncol Rev.* 2016;10(2):311.
5. Odenike O, Onida F, Padron E. Myelodysplastic syndromes and myelodysplastic/myeloproliferative neoplasms: An update on risk stratification, molecular genetics, and therapeutic approaches including allogeneic hematopoietic stem cell transplantation. *Am Soc Clin Oncol Educ Book.* 2015:e398–412.
6. Gill H, Leung AY, Kwong YL. Molecular and cellular mechanisms of myelodysplastic syndrome: Implications on targeted therapy. *Int J Mol Sci.* 2016;17(4):440.
7. Glenthøj A, Ørskov AD, Hansen JW, Hadrup SR, O'Connell C, Grønbæk K. Immune mechanisms in myelodysplastic syndrome. *Int J Mol Sci.* 2016;17(6):E944.
8. PDQ Adult Treatment Editorial Board. *Myelodysplastic Syndromes Treatment (PDQ®): Health Professional Version.* PDQ Cancer Information Summaries [Internet]. Bethesda, MD: National Cancer Institute (US); 2002–2015.

18
Acute Myeloid Leukemia

Rina Kansal MD

18.1 Definition

Myeloid neoplasms are classified into eight major entities: (i) myeloproliferative neoplasms; (ii) mastocytosis; (iii) myeloid/lymphoid neoplasms with eosinophilia and rearrangement of *PDGFRA, PDGFRB*, or *FGFR1* or with *PCM1-JAK2;* (iv) Myelodisplastic/myeloproliferative neoplasms; (v) Myelodisplastic syndromes/neoplasms; (vi) myeloid neoplasms with germline predisposition; (vii) acute myeloid leukemia and related precursor neoplasms; and (viii) acute leukemias of ambiguous lineage [1–3].

Acute myeloid leukemia (AML) is the most common malignant myeloid neoplasms and results from a clonal proliferation of progenitor myeloid cells ("blasts" or leukemic cells) that concurrently lose their normal ability to differentiate into mature myeloid cells. Previously, a diagnosis of AML required 20% or more myeloid blasts in the peripheral blood or bone marrow. In the 2008 World Health Organization (WHO) classification, the role of submicroscopic gene mutations in addition to that of chromosomal abnormalities for AML was recognized [2]. Further advances in our knowledge of the genomic architecture of hematopoietic neoplasms, especially for myeloid malignancies following the large-scale sequencing efforts in human malignant tumors after the completion of the Human Genome Project, led to the 2016 WHO classification (the revised fourth edition) for hematopoietic neoplasms, including for AML [1,3].

18.2 Biology

The majority of AML occurs *de novo*. Few are secondary to a myeloid neoplasm such as myelodisplastic syndrome (MDS) or a myeloproliferative neoplasm (MPN) or may occur after cytotoxic or other leukemogenic therapy (therapy-related) and are differentiated by clinical history.

AML is a disease with extreme genetic heterogeneity. Cytogenetic abnormalities form the basis for the three main prognostic categories in AML, with the largest cytogenetic group being that with normal cytogenetics

(CN-AML), comprising about 45% of adult AML cases and with intermediate prognostic risk. The molecular basis of AML requires cooperation among genetic mutations of at least five different classes. In adult CN-AML, the most common recurrent molecular abnormalities are mutations in *NPM1* (45%–60%), *FLT3* internal tandem duplication (ITD; 28%–34%), *DNMT3A* (30%–37%), *IDH1* and *IDH2* (25%–30%), *ASXL1* (5%–12%), *TET2* (9%–23%), and *RUNX1* (8%–16%) [4].

18.3 Epidemiology

AML occurs far more commonly in adults and in men compared with women (the male to female ratio is 3:2), with 67 years as the median age at diagnosis. The disease has an increasing incidence in the United States (with an average rise of 3.4% each year for the last 10 years), due primarily to longer survival, including after treatment for prior non-hematologic malignancies. In 2016, 19,950 new cases of AML, or 4.1 new cases per 100,000 men and women were expected to occur in the United States, with estimated 10,340 deaths due to AML. Among the new cases of AML, only 5.3% of cases occur in ages below 20 years, with infancy (less than 1 year of age) being the most common age group among pediatric AML cases.

18.4 Pathogenesis

Tumorigenesis of AML is a multistep genetic process, with a complex clonal structure that undergoes considerable evolution before the appearance of clinical signs. AML with shorter latency is usually associated with topoisomerase II inhibitor therapy and has abnormalities of *KMT2A* or *RUNX1*. AML with longer latency is more commonly associated with alkylating chemotherapy, deletions of chromosomes 5 and 7, complex karyotype, and a myelodisplastic phase. Clonal cell populations appear to evolve during the latency period, and several clones may be present at the diagnosis of AML. Clinically, increased clonal heterogeneity at diagnosis contributes to resistance to chemotherapy in any cancer, including in AML, with relapse often occurring due to clonal expansion of a previously present resistant subclone.

Clonal abnormalities may occur in normal individuals without evidence of a blood disorder. *DNMT3A* R882 mutations, which occur often in CN-AML, were most commonly identified in normal individuals—even in ages under 25 years—but with normally increased prevalence with age and having a low variant allele frequency of <3%. Because *NPM1* mutations co-occur frequently with *DNMT3A* mutations in AML, *NPM1* mutations appear to be closely related to the evolution of AML, with *DNMT3A* mutations possibly allowing the *NPM1* mutant clones to be founded and expand toward AML. In addition, the presence of clonal hematopoiesis in normal individuals also has implications for the interpretation of mutational analysis findings in the clinical management of AML. Clonal hematopoiesis may also be detected in young, asymptomatic carriers of *RUNX1* mutations, providing a possible biomarker in these families with high risk for development of AML.

Molecular modeling of complete genomic structure in 1,540 AML patients recently defined at least 11 molecular and clinically distinct classes of AML (AML with *NPM1* mutations; chromatin/RNA splicing mutations [in *RUNX1, ASXL1, BCOR, STAG2, EZH2, SRSF2, SF3B1, U2AF1, ZRSR2,* or *MLL* PTD]; *TP53* mutations/complex karyotype, inv[16] or t[16;16]; biallelic *CEBPA* mutations, t[15;17], t[8;21]; AML with *MLL* fusion genes, inv[3] or t[3;3], t[6;9]; and, provisionally, AML with *IDH2* R172 mutations) [4].

18.5 Clinical features

The clinical features of AML are varied, with symptoms usually related to low or high blood counts and often with infection. Symptoms due to tissue infiltration may be present. In the absence of treatment, AML is fatal due to bone marrow failure. Acute promyelocytic leukemia is frequently associated with disseminated intravascular coagulation. AML with t(8;21)(q22;q22) may have tumor manifestations, with the bone marrow at diagnosis showing less than 20% blasts. AML with t(1;22)(p13;q13) is a *de novo* AML that presents only in infants and young children less than 3 years of age, with most cases occurring in infancy and with patients presenting with marked hepatosplenomegaly.

Hematopoietic malignancies are usually considered to be sporadic, with very few recognized familial cases. With increasing evaluation of molecular genetics in myeloid neoplasms, there is now increasing appreciation of the need to assess familial predisposition for MDS and AML, leading to a new category "myeloid neoplasms with germline predisposition" in the 2016 WHO classification [1,3]. Clinical testing is available for familial myeloid disorders, including for the autosomal-dominant familial platelet disorder with thrombocytopenia and germline *RUNX1* mutation with a propensity to develop to AML and for familial AML with biallelic mutated *CEBPA*.

18.6 Diagnosis

Diagnosis of AML requires the presence of 20% or more myeloid blasts (blasts with a myeloid immunophenotype) in peripheral blood or bone marrow or fewer than 20% blasts in the presence of recurrent cytogenetic abnormalities specific for AML. Table 18.1 shows the 2016 WHO classification according to the revised fourth edition as is now recommended to classify AML for diagnostic purposes. In addition to cytogenetic analysis, molecular analysis for mutations is now also required up-front in the pathologic diagnostic workup of AML. Clinical history and cytogenetics are still required for precise diagnostic classification; if a newly diagnosed AML patient has Down

Table 18.1 The 2016 World Health Organization Classification of Acute Myeloid Leukemia

Acute myeloid leukemia (AML) with recurrent genetic abnormalities
AML with t(8;21)(q22;q22); *RUNX1-RUNX1T1*
AML with inv(16)(p13.1q22) or t(16;16)(p13.1;q22); *CBFB-MYH11*
Acute promyelocytic leukemia with t(15;17)(q22;q12); *PML-RARA*
AML with t(9;11)(p22;q23); *MLLT3-MLL*
AML with t(6;9)(p23;q34); *DEK-NUP214*
AML with inv(3)(q21q26.2) or t(3;3)(q21;q26.2); *GATA2, MECOM(EVI1)*
AML (megakaryoblastic) with t(1;22)(p13;q13); *RBM15-MKL1*
Provisional entity: *Acute myeloid leukemia* with *BCR-ABL1*
AML with gene mutations
AML with mutated *NPM1*
AML with biallelic mutations of *CEBPA*
Provisional entity: *Acute myeloid leukemia with mutated RUNX1*
AML with myelodysplasia-related changes
Therapy-related myeloid neoplasms
AML, not otherwise specified
AML with minimal differentiation
AML without maturation
AML with maturation
Acute myelomonocytic leukemia
Acute monoblastic and monocytic leukemia
Acute erythroid leukemia
Acute megakaryoblastic leukemia
Acute basophilic leukemia
Acute panmyelosis with myelofibrosis
Myeloid sarcoma
Myeloid proliferations related to Down syndrome
Transient abnormal myelopoiesis associated with Down syndrome
Myeloid leukemia associated with Down syndrome

syndrome or a prior history of cytotoxic or leukemogenic therapy, the classification is "myeloid proliferation of Down syndrome" or "therapy-related AML," respectively, and if, with cytogenetics, a recurrent genetic abnormality is found, then the diagnosis is "AML with recurrent genetic abnormality." However, if *NPM1* is mutant, *CEBPA* is biallelic mutant, or *RUNX1* is mutant, then regardless of any morphologic dysplasia that may be present, the diagnosis is "AML with mutated *NPM1*," "AML with mutated *CEBPA*," or "AML with mutated *RUNX1*," respectively, in the absence of a history of MDS or MPN and in the absence of cytogenetic abnormalities specific for MDS as described by the WHO classification because those cytogenetic abnormalities are sufficient to diagnose AML with MDS-related changes in the presence of 20% or more peripheral blood or marrow blasts. For any of the AML entities with specific mutated genes (*NPM1*, *CEBPA*, or *RUNX1*), if there is a history of prior MDS or MPN, the diagnosis is "AML with MDS-related changes," and if there is Down syndrome or prior leukemogenic therapy, the classification would be related to the clinical condition (Down syndrome or therapy-related) and not to an AML with gene mutations. The presence of an MDS-related cytogenetic abnormality, as described in the 2008 WHO fourth edition, excluding del(9q) if *NPM1* is mutant, also classifies as an AML with MDS-related changes, instead of AML with mutated *NPM1*, with mutated *CEBPA*, or with mutated *RUNX1* [5].

18.7 Treatment

Complete remission of AML is defined as a marrow with fewer than 5% blasts, a neutrophil count greater than 1,000, and a platelet count greater than 100,000. Patients are typically treated with cytarabine and anthracycline-based induction chemotherapy. The second phase of therapy aims to prolong the remission that, once achieved for 3 years, decreases the likelihood of relapse to less than 10%. About 20% of patients never achieve remission (primary refractory) and about 40% relapse.

Patients in remission may have minimal residual disease (MRD), which leads to relapse. Detection of MRD can alter therapy to prevent relapse or may avoid further therapy in the absence of MRD. Molecular approaches to detect MRD in AML include quantitative polymerase chain reaction (PCR), multi-gene mutational profiling, and deep sequencing for transcripts of fusion genes (e.g., *PML–RARA*) and mutations in the *NPM1, DNMT3A, IDH1, IDH2*, and *RUNX1* genes [6].

Although the current pace of advances in characterizing the genomic architecture of AML far exceeds that of available therapies for AML, MRD detection in AML is also clinically significant due to developed therapeutic agents targeting specific genetic mutations in AML. Therapeutic targeting of *FLT3* mutations is desirable because *FLT3* mutations are the single most important molecular prognostic marker, conferring a poor prognosis in AML, and several agents have been developed targeting both the *FLT3*-ITD and the point mutation in codon 835 of the second tyrosine kinase domain of the *FLT3* gene. The addition of midostaurin, an agent that targets *FLT3*-ITD and *FLT3*-TKD (tyrosine kinase domain), to standard chemotherapy and for 1 year of maintenance therapy significantly improved event-free and overall survival in patients whose leukemic blasts had a TKD or ITD (low or high *FLT3* mutation burden), indicating that *FLT3* inhibitors will soon become mandatory for the standard of care in AML with *FLT3* mutations.

Further, for clinical management of patients and families with a predisposition to familial MDS/AML, recommendations include that all carriers of such predisposing genetic mutations undergo a baseline bone marrow biopsy, twice annual physical examinations, and complete blood counts with differential counts. Although 100% of carriers with mutated *CEBPA* develop AML, albeit with long latency periods, the development of AML in the familial platelet disorder with *RUNX1* mutations is variable (20%–60%). The detection of germline variants that are similar to variants present in patients with MDS/AML should prompt the screening of family members.

18.8 Prognosis

At the time of clinical presentation, prognosis depends upon patient age, pretreatment cytogenetics, white blood cell counts, and mutational profiling. Increasing age is associated with an inferior outcome. Cytogenetic abnormalities at presentation were recognized as the most important prognostic variable for AML, with three main risk groups (favorable, intermediate, and poor), with the intermediate risk—cytogenetically normal AML (CN-AML)—comprising the largest group. As per the National Comprehensive Cancer Network (NCCN) guidelines for AML, *the favorable risk group* includes AML with recurrent translocations, including t(8;21)(q22;q22), inv(16)(p13.1q22) or t(16;16)(p13.1;q22), and t(15:17)(q22;q12). *The poor risk group* includes the t(6;9) and t(3;3) translocations, inv(3), AML with complex karyotypes (>3clonal chromosomal abnormalities), monosomal karyotypes (at least two autosomal monosomies or one autosomal monosomy and one structural chromosomal abnormality), -5, 5q-, -7, 7q-, 11q23—non t(9;11), and t(9;22). Clinically, the poor risk group includes therapy-related AML and those secondary to MDS or MPNs. *The intermediate risk group* includes CN-AML, +8 alone, t(9;11), and other non-defined clonal aberrations.

Molecular abnormalities further redefine the aforementioned cytogenetic risk groups as follows: (i) For t(8;21), inv(16), and t(16;16), the favorable risk worsens to intermediate with the presence of a *KIT* mutation, the most common of which, in codon D816, shows no response to the tyrosine kinase inhibitor, imatinib. (ii) The presence of *NPM1* mutation in the absence of *FLT3*-ITD or an isolated biallelic *CEBPA* mutation in CN-AML improves the risk from intermediate to favorable. (iii) *FLT3*-ITD mutation with CN-AML worsens the risk from intermediate to poor. (iv) Finally, co-occurring gene mutations in AML further affect the prognosis of singly mutated genes. Importantly, *FLT3*-ITD, when it co-occurs with both mutant *NPM1* and mutant *DNMT3A,* has the worst effect on patient survival, as compared with the effect of *FLT3*-ITD alone, or in single combination with either *NPM1* or *DNMT3A*, and regardless of the mutant to wild type *FLT3*-ITD ratio [7,8].

References

1. Arber DA. Principles of classification of myeloid neoplasms. In Jaffe ES, Arber DA, Campo E, Harris NL, Quintanilla-Martinez L, editors. *Hematopathology.* Elsevier, Amsterdam, The Netherlands, 2017. p. 785–92.
2. Swerdlow SH, Campo E, Harris NL, Jaffe ES, Pileri SA, Stein H, Thiele J, Vardiman JW (eds.): *WHO Classification of tumours of hematopoietic and lymphoid tissues.* IARC: Lyon, Amsterdam, The Netherlands, 2008.
3. Arber DA. Acute myeloid leukemia. In Jaffe ES, Arber DA, Campo E, Harris NL, Quintanilla-Martinez L, editors. *Hematopathology.* Elsevier, 2017. p. 817–45.
4. Papaemmanuil E, Gerstung M, Bullinger L, et al. Genomic classification and prognosis in acute myeloid leukemia. *NEJM* 2016;374:2209–21.
5. Dohner H, Estey EH, Amadori S, et al. Diagnosis and management of acute myeloid leukemia in adults: Recommendations from an international expert panel, on behalf of the European LeukemiaNet. *Blood.* 2010;115:453–74.
6. Falini B, Sportoletti P, Brunetti L, Martelli MP. Perspectives for therapeutic targeting of gene mutations in acute myeloid leukaemia with normal cytogenetics. *Br J Haematol.* 2015;170:305–22.
7. National Comprehensive Cancer Network. *Guidelines Version 2.2016 for acute myeloid leukemia.* www.nccn.org. Accessed September 25, 2016.
8. Kansal R. Acute myeloid leukemia in the era of precision medicine: Recent advances in diagnostic classification and risk stratification. *Canc Biol Med.* 2016;13:41–54.

19 Blastic Plasmacytoid Dendritic Cell Neoplasms (BPDCN)

19.1 Definition

Grouped along with myeloproliferative neoplasms (MPN), mastocytosis, myeloid/lymphoid neoplasms with eosinophilia and rearrangements, myelodisplastic/myeloproliferative neoplasms (MDS/MPN), myelodisplastic syndrome (MDS), acute myeloid leukemia (AML), acute leukemia of ambiguous lineage (ALAL), B-lymphoblastic leukemia/lymphoma, and T-lymphoblastic leukemia/lymphoma under the myeloid neoplasms and acute leukemia category (see Chapter 13), blastic plasmacytoid dendritic cell neoplasms (BPDCN, formerly known as natural killer cell leukemias/lymphomas) comprise two clinically and pathologically distinct variants: dermatopathic and leukemic [1,2].

Forming plasmacytoid dendritic cell tumors (so-called "blastic plasmacytoid dendritic cell neoplasms," the BPDCN dermatopathic variant shows a distinctive cutaneous and bone marrow tropism with rapid systemic dissemination. Forming nodular aggregates of clonally expanded plasmacytoid dendritic cells in the lymph nodes, skin, and bone marrow (so-called "mature plasmacytoid dendritic cell proliferation associated with myeloid neoplasms"), the BPDCN leukemic variant is rare, affecting predominantly males, and almost invariably is associated with a myeloid neoplasm (e.g., chronic myelomonocytic leukemia or other myeloid proliferations with monocytic differentiation) [3].

19.2 Biology

Typically distributed close to high endothelial venules of the lymph nodes as clusters or dispersed cells and scattered in other lymphoid tissues (e.g., tonsils, spleen, and mucosa-associated lymphoid tissue) and the bone marrow, plasmacytoid dendritic cells are medium-sized with a round-ovoid or elongated nucleus, fine chromatin, and small nucleoli; they have a low proliferation index (<10% Ki-67).

Circulating plasmacytoid dendritic cells are rare (0.01%–0.5%) and tend to decrease with age. During an ongoing immune reaction, a significant number of plasmacytoid dendritic cells accumulate in the lymph nodes; whereas during inflammatory or neoplastic diseases, recruitment of plasmacytoid dendritic cells to non-lymphoid tissues takes place.

As highly specialized circulating cells of the innate immune system, plasmacytoid dendritic cells are capable of producing high levels of type I interferon (IFN-I) and differentiating clonally into antigen-presenting dendritic cells upon stimulation. Thus, by bridging the innate and acquired components of the immune responses, plasmacytoid dendritic cells play important roles in defense against pathogens, cancer, and autoimmunity.

Besides type I IFN-producing plasmacytoid dendritic cells, two myeloid dendritic cell populations (i.e., CD1c+[BDCA1+] dendritic cells and CD141+[BDCA3+] dendritic cells) are identified in the blood and peripheral tissues. Despite originating possibly from a common macrophage/dendritic cell progenitor, plasmacytoid dendritic cells in the peripheral blood are CD11c–/CD123+/CD303+, while myeloid dendritic cells are CD11c+/CD123–/CD303–. Together with myeloid dendritic cells, type I IFN-producing plasmacytoid dendritic cells complete the contingent of human dendritic cells.

Of the two main clinicopathological entities within BPDCN, "blastic plasmacytoid dendritic cell neoplasm" originates from plasmacytoid dendritic cell precursors; while "mature plasmacytoid dendritic cell proliferation associated with myeloid neoplasms" evolves from fully differentiated plasmacytoid dendritic cells and is regularly associated with another myeloid neoplasm.

The cutaneous tropism of BPDCN tumor cells is highlighted by their expression of antigens (e.g., CLA and CD56) that favor skin migration and by their expression of chemokine binding cognate receptors (e.g., CXCR3, CXCR4, CCR6, and CCR7). Proliferation of malignant cells in BPDCN contributes to the reduction in red blood cell, white blood cell, and platelet counts [3].

19.3 Epidemiology

BPDCN is a rare, aggressive hematologic malignancy, accounting for 0.44% of all hematological malignancies, <1% of acute leukemias, 0.7% of cutaneous lymphomas, and 6.3% of the NK-cell lineage malignancies. It typically occurs in the skin of older adults (median diagnostic age of 67 years, range of 8–103 years) and displays a male predilection (3:1). About 5% of BPDCN occur in patients <10 years of age.

19.4 Pathogenesis

Risk factors for BPDCN include viral infection (e.g., human T-cell lymphotropic virus), prior chemotherapy, and other myeloid neoplasms.

Molecularly, BPDCN is linked to multiple karyotypic abnormalities (e.g., chromosomes 5q [72%], 12p [64%], 13q [64%], 6q [50%], 15q [43%], and monosomy 9 [28%]) and genetic mutations involving 9p21.3 (CDKN2A/CDKN2B, 50%), 13q13.1-q14.3 (RB1, 43%), 12p13.2-p13.1 (CDKN1B, 64%), 13q11-q12 (LATS2), TET2 (36%–80%), ASXL1 (32%), NPM1 (20%), NRAS (20%), IKZF1 (20%), IKZF1-3 (20%), ZEB2 (16%), TP53 (14%), HOXB9 (4%), and UBE2G2 (4%) [4,5].

19.5 Clinical features

Clinically, the BPDCN dermatopathic variant has a deceptive indolent onset dominated by nonpruritic skin lesions (e.g., multiple nodules [73% of cases], bruise-like infiltrates [12%], or plaques of variable size [a few mm to several cm] and color [erythematous, reddish, or bluish]) followed by rapid leukemic dissemination.

The BPDCN leukemic variant is characterized by an elevated white blood cell count, circulating blasts, and massive bone marrow infiltration along with a localized or generalized lymphadenopathy (40% of cases), splenomegaly (25%), hepatomegaly (16%), multiple skin lesions (multiple erythematous macules or papules), anemia, and thrombocytopenia. Rare cases of leukemic variants may be associated with another myeloid neoplasm (e.g., acute myeloid leukemia, chronic myeloid leukemia, chronic myelomonocytic leukemia, subacute myelomonocytic leukemia, and myelodysplasia).

19.6 Diagnosis

Patients suspected of BPDCN (particularly the elderly with nonspecific persistent skin lesions refractory to treatment) should undergo blood cell counts and a skin/bone marrow biopsy to demonstrate characteristic cell morphology and specific immunophenotype (detectable by immunohistochemistry or flow cytometry).

Blood cell counts show leukocytosis ($>10 \times 10^9$ WBC/L) in 33% of the cases, anemia (<100 g hemoglobin/L) in 63% of the cases, thrombocytopenia ($<100 \times 10^9$/L) in 64% of the cases, and neutropenia ($<1.5 \times 10^9$/L) in 48% of the cases. Lactate dehydrogenase (LDH) serum levels are increased in

55% of the cases. Blast cells are present in both peripheral blood (61% of the cases, median blast cell percentage: 39%) and bone marrow (93% of cases, median blast cell percentage: 78%) [4].

Microscopically, skin biopsy typically shows a diffuse monomorphic infiltrate of medium-sized plasmacytoid cells with scant cytoplasms, irregular nuclei, fine chromatin, at least one small nucleolus. Malignant cells usually involve the dermis with extension to the subcutaneous fat but spare the epidermis and adnexa. Mitoses are regularly present, and Ki-67-positive cells vary from 20%–80% [4].

Bone marrow biopsy may reveal small interstitial infiltrates of plasmacytoid dendritic cells (as detected by immunohistochemistry or flow cytometry), diffuse bone marrow involvement, and dysplastic changes in residual hematopoietic tissue, particularly in megakaryocytes.

Immunohistochemically, BPDCN are positive for CD4, CD56, and plasmacytoid dendritic cell markers (CD123 [interleukin 3α chain receptor], BDCA-2 [blood dendritic cell antigen or CD303], and TCL1 [T-cell leukemia 1]) but are negative for B- (CD19, CD20, and CD79a), T- (CD3 and CD5), and myelomonocytic lineage (CD13 and myeloperoxidase)-specific markers. Simultaneous expression of CD4, CD56, CD123, CD303, and TCL1 markers is observed in 46% of BPDCN cases, but the expression of four of these five markers is sufficient for a reliable diagnosis. Tumors that do not meet the criteria for BPDCN may possibly belong to "acute leukemia of ambiguous lineage" (see Chapter 20) [6].

In flow cytometry, BPDCN display a high-intensity expression of CD123 and positivity for HLA-DR, CD4, CD45RA, and CD56 but negativity for lineage-associated antigens. This immunophenotype is considered unique and virtually pathognomonic for BPDCN.

Despite accounting for fewer than 0.4% of peripheral blood mononuclear cells in BPDCN, nonmalignant plasmacytoid dendritic cells can accumulate in various pathological conditions such as autoimmune diseases, classical Hodgkin lymphoma, and carcinomas. Differential diagnoses for BPDCN include a cutaneous T-cell lymphoma with skin lesions and blood involvement (which shows CD56 negativity and disproportionate epidermotropism), an extranodal NK/T-cell lymphoma with skin lesions and the expression of the CD4+/CD56+ immunophenotype (which is Epstein–Barr virus positive), an extramedullary myeloid sarcoma (EMS, which is positive staining for lysozyme or myeloperoxidase but negative for CD56, CD123, and TCL1), non-Hodgkin lymphoma (NHL), acute myeloid leukemia (AML, CD303 negative), leukemia cutis (cutaneous manifestation of any type of leukemia, CD303 negative), melanoma, and lupus erythematosus.

19.7 Treatment

For BPDCN patients with cutaneous lesions, which often occur in early stage BPDCN without obvious systemic involvement, treatment options consist of focal radiotherapy, systemic glucocorticosteroids, or nonintensive chemotherapy regimens. This may lead to the complete resolution of cutaneous lesions but does not appear to provide a long-term benefit, with most patients relapsing within several months.

For patients with advanced-stage BPDCN, intensive chemotherapy regimens derived from the management of non-Hodgkin lymphoma (CHOP [cyclophosphamide/hydroxydaunomycin/vincristine/prednisone] or CHOP-like), acute lymphoblastic leukemia (ALL; hyper-CVAD [hyperfractionated/cyclophosphamide/vincristine/doxorubicin/dexamethasone] alternating with methotrexate and cytarabine), and AML may be utilized. CHOP-like regimens demonstrate a response rate of 86%, a median time to relapse of 9 months involving the bone marrow and central nervous system, and an overall survival rate of 25% after 24 months. More intensive, hyper-CVAD like regimens show a response rate of 90%, a median duration of response of 20 months, and a median overall survival rate of 29 months. For relapsed BPDCN, which often has a reduced response to previous chemotherapy, alternative drugs (e.g., L-asparaginase/methotrexate/dexamethasone) may be considered.

Allogeneic hematopoietic stem cell transplantation (allo-HSCT) significantly improves the rate of relapse (to 32% at 3 years if offered in first remission) [7]. It is, therefore, important for BPDCN patients to be evaluated for an allo-HCT as soon as possible and to begin searching for a donor.

Alternative therapies under investigation include the use of diphtheria toxin to target IL3 receptors on BPDCN cells for specific apoptosis and the application of surface receptors with immunotoxins (i.e., interleukin-3 receptor alpha) or inhibitors of aberrantly activated survival pathways (i.e., NF-κB).

19.8 Prognosis

As BPDCN have a relatively short response to conventional chemotherapy regimens, the long-term prognosis is poor. The median survival for most BPDCN patients is less than 18 months and that for patients with the leukemic form of the disease is 8.7 months. However, pediatric BPDCN cases have a more favorable outcome.

Unfavorable prognostic factors include low TdT expression, positivity for BDCA2/CD303, *CDKN2A/CDKN2B* deletions, and mutations in DNA methylation pathway genes.

References

1. Swerdlow SH, Campo E, Harris NL, et al. (eds). *WHO classification of tumours of haematopoietic and lymphoid tissues.* International Agency for Research on Cancer; World Health Organization. Lyon, France: IARC, 2008.
2. Arber DA, Orazi A, Hasserjian R, et al. The 2016 revision to the World Health Organization classification of myeloid neoplasms and acute leukemia. *Blood.* 2016;127(20):2391–405.
3. Facchetti F, Cigognetti M, Fisogni S, Rossi G, Lonardi S, Vermi W. Neoplasms derived from plasmacytoid dendritic cells. *Mod Pathol.* 2016;29(2):98–111.
4. Riaz W, Zhang L, Horna P, Sokol L. Blastic plasmacytoid dendritic cell neoplasm: Update on molecular biology, diagnosis, and therapy. *Canc Control.* 2014;21(4):279–89.
5. Nguyen CM, Stuart L, Skupsky H, Lee YS, Tsuchiya A, Cassarino DS. Blastic plasmacytoid dendritic cell neoplasm in the pediatric population: A case series and review of the literature. *Am J Dermatopathol.* 2015;37(12):924–8.
6. Martín-Martín L, López A, Vidriales B, et al. Classification and clinical behavior of blastic plasmacytoid dendritic cell neoplasms according to their maturation-associated immunophenotypic profile. *Oncotarget.* 2015;6(22):19204–16.
7. Kharfan-Dabaja MA, Lazarus HM, Nishihori T, Mahfouz RA, Hamadani M. Diagnostic and therapeutic advances in blastic plasmacytoid dendritic cell neoplasm: A focus on hematopoietic cell transplantation. *Biol Blood Marrow Transplant.* 2013;19(7):1006–12.

20
Acute Leukemia of Ambiguous Lineage (ALAL)

20.1 Definition

Grouped under the myeloid neoplasms and acute leukemia category (which also includes myeloproliferative neoplasms [MPN], mastocytosis, myeloid/lymphoid neoplasms with eosinophilia and rearrangement, myelodisplastic/myeloproliferative neoplasms [MDS/MPN], myelodisplastic syndrome [MDS], acute myeloid leukemia [AML], blastic plasmocytoid dendritic cell neoplasms (BPDCN), B-lymphoblastic leukemia/lymphoma, and T-lymphoblastic leukemia/lymphoma), acute leukemia of ambiguous lineage (ALAL) consists of (i) acute undifferentiated leukemia, (ii) mixed phenotype acute leukemia (MPAL) with t(9;22)(q34.1;q11.2)—*BCR-ABL1,* (iii) MPAL with t(v;11q23.3)—*KMT2A* rearranged, (iv) MPAL-B/myeloid not otherwise specified (NOS), and (v) MPAL-T/myeloid NOS [1,2].

Depending upon whether they are of myeloid or lymphoid lineage, acute leukemias are differentiated into acute myeloid leukemia (AML) and acute lymphoid leukemia (ALL). However, about 4% of acute leukemias show biphenotypic characteristics, and their primary stem cell defect cannot be determined conclusively. These tumors are thus referred to as ALAL [2].

20.2 Biology

Resulting from neoplastic proliferation of hematopoietic stem cells and accumulation of blasts and immature cells in the bone marrow, acute leukemias are classified on the basis of their morphology, immunophenotyping, chromosomal abnormalities, and specific molecular genetic features. Although most acute leukemias fall into the myeloid (i.e., AML) or B- or T-lymphoid (i.e., ALL) lineages, some leukemias with uncertain primary stem cell defects are assigned as MPAL, biphenotypic acute leukemia (BAL), acute undifferentiated leukemia (AUL), bilineal acute leukemia, and hybrid leukemia.

Accounting for 4% of acute leukemias, ALAL is a rare complex identity with no clear evidence of differentiation along a single lineage.

ALAL encompasses MPAL, BAL, and AUL and demonstrates heterogeneous clinical, immunophenotypic, cytogenetic, and molecular features, as well as adverse clinical outcomes.

Specifically, MPAL consists of blasts that express antigens of more than one lineage. MPAL containing two distinct populations of blasts each expressing antigens of a different lineage is called "bilineal leukemia." MPAL containing a single blast population expressing antigens of multiple lineages is called "biphenotypic acute leukemia." MPAL excludes cases with recurrent AML abnormalities t(15;17)(q22;q21), inv(16)(p13q22)/t(16;16)(p13;q22), t(8;21)(q22;q22), chronic myeloid leukemia (CML) in blast crisis, MDS-related AML, and therapy-related AML, even if they have a mixed phenotype [3,4].

20.3 Epidemiology

ALAL represents 4% of all acute leukemias, of which MPAL accounts for about 2%, MPAL with t(9;22)(q34;q11.2) (or BCR-ABL1 rearrangement) for <1%; MPAL-B/myeloid and T/myeloid types for <1%.

MPAL affects both adults and children (especially MPAL with t[v;11q23] or MLL rearrangement and MPAL-T/myeloid type) and shows a male predominance. AUL seems to occur more commonly in older adults and has a high frequency of del(5q) and trisomy 13.

20.4 Pathogenesis

Chromosomal abnormalities found in MPAL include del(1)(p32) (*STIL-TAL1* fusion, involving 1.1% of MPAL and 17% of childhood T-ALL); t(2;5)(p13;p13~15.3); t(3;11)(q13;p15) (*NUP98-IQCG* fusion gene, involving adult MPAL and T+M phenotypes); trisomy 4; monosomy 5 or deletion of 5q (−5/del[5q], involving MPAL-B+T or M+T immunophenotypes; MDS; and AML); del(6q) (involving MPAL-T+M, B+M, or M+B+T phenotypes); monosomy 7 or deletion of 7q (−7/del[7q], involving MPAL-B+M or M+T phenotypes, adult MPAL with Ph chromosome, MDS, and AML); abn(7p) (involving MPAL-M+T or M+B phenotypes); trisomy 8 (involving MPAL-M+T or B+M phenotypes); polysomy 8 (involving MPAL-B+M phenotypes, AML, and MDS); t(8;21)(q22;q22) (*RUNX1-RUNX1T1* fusion, involving MPAL-M+B phenotypes); t(9;22)(q34;q11.2) (*BCR-ABL1* fusion); t(10;11)(p15;q21) (*PICALM-MLLT10* fusion, involving MPAL-T+M or B+M phenotypes, T-ALL, and AML); t(v;11q23) (*MLL* rearranged); rearrangements of 12p (*ETV6* rearrangements, involving MPAL and BAL); del(12p) (involving child MPAL-T+M phenotypes or B+M phenotypes); t(4;12)(q12–21;p13) (*ETV6* rearrangement, involving child MPAL-B+M phenotypes); t(7;12)(q36;p13) (*MNX1-ETV6* fusion, previously termed *HLXB9-TEL* fusion, involving child MPAL-M+T phenotypes, AML); t(8;12)(q13;p13) (ETV6-NCOA2 fusion, involving MPAL-T+M antigens); t(12;21)(p13;q22) (*ETV6-RUNX1* fusion, involving 16% of child BAL); t(12;22)(p13;q12) (involving adult MPAL-B+M phenotypes); rearrangements of 14q32 (add[14][q32], t[8;14][p21;q32], and t[6;14][q25;q32], involving MPAL and BAL-T-lymphoid and myeloid-associated antigen); i(17)(q10) (involving MPAL-T+M phenotypes); trisomy 19 (involving MPAL-M+T phenotypes); trisomy 21 and polysomy 21 (involving BAL, MPAL, AML, ALL); abnormalities of chromosome X (involving MPAL-M+T or M+B phenotypes); hyperdiploid karyotypes (involving pediatric MPAL-M+B, M+T, or B+T+M phenotypes); complex karyotypes (−7/del[7q], del[6q], −5/del[5q], and −17/17p-, involving up to 50% of MPAL-B+M, T+M, and, rarely, -B+T phenotypes); normal karyotype (involving up to 36% of MPAL-B+M, T+M, B+T, or B+T+M phenotypes) [5–8].

Cytogenetic abnormalities in AUL include del(5q), trisomy 12, trisomy 13, del(20q), t(11;14)(q23;q24) (*MLL-GPHN* fusion), SET-NUP214 rearrangement, and other abnormalities (in tetrasomy 8, monosomy 7, and *IGH* rearrangements) [5].

In comparison with its adult counterpart, in which the most frequent cytogenetic abnormality is t(9;22)(q34;q11.2), resulting in Ph chromosomes, pediatric MPAL often possesses t(v;11q23), involving *MLL* rearrangement (40% of cases), abnormalities of 1p (13.3%) and 5p (13.3%), and t(7;12)(q36;p12) (13.3%).

20.5 Clinical features

Clinical symptoms associated with ALAL range from asthenia, pallor, fever, dizziness, agitation/irritability, incoordination, shortness of breath, easy bruising, excessive/prolonged menstrual bleeding, bleeding/enlarged gums, coagulation disorders, neurological disorders, severe thrombocytopenia, weakness, fatigue, palpitations, bone/chest/abdominal/urination pain, headache, enlarged liver/spleen, to weight loss.

20.6 Diagnosis

Conventional techniques for diagnosis of acute leukemia and discrimination between ALL and AML consist of morphologic, immunohistochemical, and immunological approaches, which are increasingly supplemented by gene and microRNA (miRNA) expression profiling analysis [6].

Flow cytometric immunophenotyping of cytoplasmic myeloperoxidase (MPO, the ultimate marker of myeloid lineage), CD19, cytoplasmic CD3, and other markers plays a vital role in the diagnosis of ALAL. The scoring system of the European Group for the Immunologic Characterization of Leukemias (EGICL) on the basis of degree of lineage specificity of each antigen is highly informative for the diagnosis of biphenotypic leukemia, which requires two or more points in two categories (Table 20.1).

MPAL, with t(9;22)(q34;q11.2) or BCR-ABL1 rearrangement, often contains myeloblasts and precursor B cell lymphoblasts, occasionally myeloblasts and precursor T-cell lymphoblasts, and rarely three components (myeloblasts, precursor B-cell lymphoblasts, and precursor T-cell lymphoblasts).

MPAL with t(v;11q23) or MLL rearrangement tends to have immunophenotypes of CD19+, CD15+, CD20-, CD10-, and HLA-DR+, with lymphoblasts being generally negative for all other myeloid antigens and CD24. Rare cases of MPAL with t(v;11q23) may show a mature immunophenotype (i.e., λ sIg+, CD19+, CD10-, TdT-, and CD34-), with no evidence of a *c-myc* rearrangement.

MPAL-B/myeloid and T/myeloid may be of biphenotype or mixed lineage. BAL demonstrates precursor B-cell and myeloid blasts or precursor T-cell and myeloid blasts. MPAL NOS demonstrates evidence of both precursor B-cell and precursor T-cell lineages (biphenotypic or mixed lineages).

AUL typically expresses no more than one surface membrane antigen of any given lineage and lacks T-cell-, myeloid-, B-lineage-specific markers (i.e., CD3, myeloperoxidase-MPO, cCD22, cCD79a, and strong CD19), and other lineage-specific markers (e.g., those for erythroid precursors,

Table 20.1 The EGICL Scoring System for Diagnosis of Biphenotypic Leukemia

Score	B-Lymphoid	T-Lymphoid	Myeloid
2	cyCD79a	CD3(surface/cy)	MPO
	cyIgM	TCR	
	cyCD22		
1	CD19	CD2	CD117
	CD20	CD5	CD13
	CD10	CD8	CD33
		CD10	CD65
0.5	TdT	TdT	CD14
	CD24	CD7	CD15
		CD1a	CD64

history of autoimmune diseases (e.g., autoimmune thyroid diseases, Graves' disease, or Hashimoto's thyroiditis).

Molecularly, B-ALL often contains alterations in chromosome 21q (e.g., trisomy, tetrasomy, and intrachromosomal amplification of AML1 gene [21q22]); cytogenetic aberrations (e.g., hyperdiploidy, cryptic t[12;21] [TEL-AML1 fusion, 25.4%], t[1;19][q23;p13] [E2A-PBX (PBX1) fusion, 4.8%], t[9;22][q34;q11] [BCR-ABL fusion, 1.6%], t[4;11][q21;q23] [MLL-AF4 fusion, 1.6%], and t[8;14][q24;q32] [IGH-MYC fusion]); clonal rearrangements of the immunoglobulin heavy chain genes; and occasionally the light chain genes. B-ALL with t(9;22), also known as Philadelphia+ (Ph+) ALL, is noted for its production of the BCR–ABL1 fusion protein and active kinase signaling. Further, activating mutations of the JAK proteins and cytokine receptor-like factor 2 (CRLF2) overexpression leading to activation of the JAK–STAT signaling may be responsible for the similarity in kinase signaling profiles observed in Ph+ ALL. Interestingly, frequencies of IKZF1 mutation/deletion or CRLF2 mutation with overexpression in childhood B-ALL (10%–30% or 7%–15%, respectively) differ slightly from those in adult B-ALL (15%–18% or 15%, respectively) [4,5].

T-ALL may harbor cytogenetic abnormalities, such as translocations of the T-cell receptor gene (TCRα, TCRβ, TCRγ, and TCRδ) with Lim only domain 2 (LMO2, 15%), T-cell acute leukemia 1 (TAL1, 11%), T-cell leukemia homeobox 1 (TLX1, 25%), and TCR-T-cell leukemia/lymphoma 1A (TCL1A). Given that Notch1 protein interacts directly with a myriad of pathways and players, including c-myc, PI3K/Akt signaling, PTEN, Fbxw7, and NF-κB, it is significant that point mutations that stabilize the Notch1 protein occur in more than 50% of T-ALL cases. Indeed, T-ALL containing t(7;9)(q34;q34.3) translocations often expresses a truncated NOTCH1 mRNA under the control of the TCRB promoter. Childhood T-ALL often contains mutations in NOTCH1 (>50%), FBXW7 (9%–16%), and PTEN (8%), which are rarely observed in adult T-ALL [6–8].

21.5 Clinical features

Clinical symptoms of ALL (B-ALL and T-ALL) result mainly from a reduced production of functional blood cells and include fever, cough, vomiting, generalized weakness, anemia (pallor, tachycardia, fatigue, and headache), dizziness, frequent infection, loss of appetite, excessive bruising, chest/bone/joint pain, shortness of breath, enlarged lymph nodes/liver/spleen, pitting edema (swelling) in the lower limbs and abdomen, petechiae (tiny red spots or lines in the skin due to low platelet levels), and weight loss.

21.6 Diagnosis

Morphological evaluation and immunophenotyping are fundamental to the diagnosis of ALL and to discrimination between B-ALL and T-ALL. Cytogenetic analysis of specific chromosomal abnormalities provides additional support for subclassification.

Microscopic examination of peripheral blood and bone marrow aspirate from ALL patients reveals small- to medium-sized cells with scant basophilic cytoplasm, a high nuclear-to-cytoplasmic ratio, round to oval nuclei, smudged or coarsely reticular chromatin, and variably prominent nucleoli. Bone marrow biopsy shows homogenous lymphoblasts (which may constitute 20% of the cells in acute disease) of intermediate size with minimal cytoplasm, round or convoluted nuclei, dispersed or stippled chromatin, inconspicuous or prominent nucleoli, and brisk mitotic activity.

In combination with morphologic features, immunophenotyping helps confirm the diagnosis and determine the lineage background of ALL (B-ALL and T-ALL). Specifically, B-ALL is commonly positive for CD19, cCD22, cCD79a, PAX5, CD10, sCD22, CD24, and TdT and variably positive for CD20, CD34, CD45, CD13, CD33, and sIgM. T-ALL is commonly positive for cCD3, CD7, and TdT but variably positive for CD1a, CD2, sCD3, CD4, CD5, CD8, CD10, CD34, CD99, CD19, CD33, CD79a, CD117, and CD56.

If aspirate is unavailable, immunohistochemistry may be employed for immunophenotypic characterization. Immunohistochemically, ALL is positive for periodic acid Schiff (PAS) stain, variably positive for nonspecific esterase and Sudan Black B, and negative for myeloperoxidase (MPO). In addition, B-ALL is positive for B-cell markers (eh, CD19, CD79a, and CD22), CD10 (common acute lymphocytic leukemia antigen and CALLA), CD24, PAX5, TdT, and is variably positive for CD20, CD34, CD45, and CD99. The "pro-B" stage (CD10-, CD19+, CD79a+, CD22+, nuclear TdT+), "common" stage (CD10+), and "late pre-B" stage (CD20+, cytoplasmic heavy chain+) of B-ALL can be also determined immunohistochemically. On the contrary, T-ALL is positive for TdT (intracellularly), CD2, CD3 (most specific), CD4, CD5, CD7, and CD8. Given the morphological similarity between T-ALL and B-ALL, use of T-cell markers (e.g., CD1a, CD2, CD3, CD4, CD5, CD7, and CD8) is critical for the verification of T-ALL [4,5].

Molecularly, most cases of T-ALL do not have rearrangement in the TCR gene, while 50%–70% of T-ALL patients contain karyotype abnormalities (e.g., translocations between TCRα or δ on the short arm of chromosome 14 [14q11.2], TCRβ on the long arm of chromosome 7 [7q35], or TCRγ on the short arm of chromosome 7 [7p14-15] and various partner genes) [6–8].

Differential diagnoses for extramedullary B-ALL (TdT positive, MPO positive, surface Ig positive, cyclin D1, and CD5 negative) include aggressive mature B-cell lymphomas (blastoid mantle cell lymphoma, Burkitt lymphoma, and double-hit and other gray zone lymphomas), Ewing family tumors, and myeloid leukemias. Positivity of CD10 and/or CD34, presence of a tight cluster of lymphoblasts, and absence of surface CD3, CD2, or CD5 in flow cytometry help differentiate T-ALL from thymoma.

21.7 Treatment

Acute leukemia in children (80%) and adults (20%–40%) is curable with standard treatments, which consist of multi-agent chemotherapy regimens (e.g., CVAD and rituximab) for B-ALL and the addition of steroids for T-ALL. Older or less fit patients respond to chemotherapy initially but often experience relapse.

Intrathecal chemotherapy helps prevent CNS relapses of T-ALL and may negate the need for cranial irradiation. Mediastinal irradiation may reduce mediastinal relapses of T-ALL. Allogeneic or autologous stem cell transplantation/bone marrow transplantation may be considered for younger/fitter patients with T-ALL if chemotherapy does not cure the disease.

Molecular analysis is playing an increasingly important role in the management of ALL. Classifying cases as having t(9;22) enables targeted therapy with imatinib. Identification of Ph-like B-ALL paves the way for treatment with other specific kinase inhibitors. Recognition of hypermethylation in ALL may facilitate potential treatment with methylation inhibitors.

21.8 Prognosis

Due largely to risk stratification and the resulting ability to tailor therapy appropriately, survival of patients with ALL has improved dramatically over the years. Currently, B-ALL has a 5-year disease-free survival rate of 85% for children and 45%–72% for adults. T-ALL has a relatively poor prognosis compared to B-ALL. Intensified chemotherapy protocols improve the prospect of T-ALL, with a cure rate of 75% in children and 50% in adults.

References

1. Swerdlow SH, Campo E, Harris NL, et al. (eds). *WHO classification of tumours of haematopoietic and lymphoid tissues.* International Agency for Research on Cancer; World Health Organization. Lyon, France: IARC, 2008.
2. Arber DA, Orazi A, Hasserjian R, et al. The 2016 revision to the World Health Organization classification of myeloid neoplasms and acute leukemia. *Blood.* 2016;127(20):2391–405.
3. McGregor S, McNeer J, Gurbuxani S. Beyond the 2008 World Health Organization classification: The role of the hematopathology laboratory in the diagnosis and management of acute lymphoblastic leukemia. *Semin Diagn Pathol.* 2012;29(1):2–11.
4. Geethakumari PR, Hoffmann MS, Pemmaraju N, et al. Extramedullary B lymphoblastic leukemia/lymphoma (B-ALL/B-LBL): A diagnostic challenge. *Clin Lymphoma Myeloma Leuk.* 2014;14(4):e115–8.
5. Jung SI, Cho HS, Lee CH, Jung BC. A case of adult B lymphoblastic leukemia with ider(9)(q10)t(9;22)(q34;q11.2) and der(19)t(1;19)(q23;p13.3). *Kor J Lab Med.* 2010;30(6):585–90.
6. Van Vlierberghe P, Ferrando A. The molecular basis of T cell acute lymphoblastic leukemia. *J Clin Invest.* 2012;122:3398–406.
7. Sugimoto KJ, Shimada A, Wakabayashi M, et al. T-cell lymphoblastic leukemia/lymphoma with t(7;14)(p15;q32) [TCRγ-TCL1A translocation]: A case report and a review of the literature. *Int J Clin Exp Pathol.* 2014;7(5):2615–23.
8. You MJ, Medeiros LJ, Hsi ED. T-lymphoblastic leukemia/lymphoma. *Am J Clin Pathol.* 2015;144(3):411–22.

22 Mature B-Cell Neoplasms

22.1 Definition

Tumors of the hematopoietic and lymphoreticular systems consist of two main categories: (i) myeloid neoplasms and acute leukemia (representing 35% of all hematopoietic and lymphoreticular neoplasms) and (ii) mature lymphoid, histiocytic, and dendritic neoplasms (representing 65% of all hematopoietic and lymphoreticular neoplasms) [1,2].

Mature lymphoid, histiocytic, and dendritic neoplasms encompass five groups: (i) mature B-cell neoplasms, (ii) mature T and natural killer (NK) neoplasms, (iii) Hodgkin lymphoma, (iv) post-transplant lymphoproliferative disorders (PTLD), and (v) histiocytic and dendritic cell neoplasms. Of these, mature B-cell neoplasms and mature T and NK neoplasms (see Chapter 24) are collectively referred to as non-Hodgkin lymphoma (NHL) [2].

Accounting for about 85% of NHL cases, mature B-cell neoplasms are subdivided into at least 45 types (i.e., chronic lymphocytic leukemia/small lymphocytic lymphoma [CLL/SLL], monoclonal B-cell lymphocytosis, B-cell prolymphocytic leukemia, splenic marginal zone lymphoma, hairy cell leukemia [splenic B-cell lymphoma/leukemia unclassifiable—splenic diffuse red pulp small B-cell lymphoma/hairy cell leukemia-variant], lymphoplasmacytic lymphoma [LPL], Waldenström macroglobulinemia, monoclonal gammopathy of undetermined significance [MGUS] IgM, μ heavy-chain disease, γ heavy-chain disease, α heavy-chain disease, monoclonal gammopathy of undetermined significance [MGUS] IgG/A, plasma cell myeloma, solitary plasmacytoma of bone, extraosseous plasmacytoma, monoclonal immunoglobulin deposition diseases, extranodal marginal zone lymphoma of mucosa-associated lymphoid tissue [MALT lymphoma], nodal marginal zone lymphoma—pediatric nodal marginal zone lymphoma, follicular lymphoma [FL], *in situ* follicular neoplasia, duodenal-type follicular lymphoma, pediatric-type follicular lymphoma—large B-cell lymphoma with IRF4 rearrangement, primary cutaneous follicle center lymphoma, mantle cell lymphoma [MCL], in situ mantle cell neoplasia, diffuse large B-cell lymphoma not otherwise specified [DLBCL NOS], germinal center B-cell type, activated B-cell type, T-cell/histiocyte-rich large B-cell lymphoma, primary DLBCL of the central nervous system [CNS], primary cutaneous DLBCL leg type,

EBV+ DLBCL NOS—EBV+ mucocutaneous ulcer, DLBCL associated with chronic inflammation, lymphomatoid granulomatosis, primary mediastinal [thymic] large B-cell lymphoma, intravascular large B-cell lymphoma, ALK+ large B-cell lymphoma, plasmablastic lymphoma, primary effusion lymphoma—HHV8+ DLBCL NOS, Burkitt lymphoma—Burkitt-like lymphoma with 11q aberration, high-grade B-cell lymphoma with *MYC* and *BCL2* and/or *BCL6* rearrangements, high-grade B-cell lymphoma NOS, and B-cell lymphoma unclassifiable with features intermediate between DLBCL and classical Hodgkin lymphoma) [2].

The most common mature B-cell neoplasms are diffuse large B-cell lymphoma (DLBCL, 37%; including primary mediastinal large B-cell lymphoma, 3%), follicular lymphoma (FL, 29%), chronic lymphocytic leukemia/small lymphocytic lymphoma [CLL/SLL, 12%), marginal zone lymphoma (MZL; MALT lymphoma, 9%; nodal marginal zone lymphoma, NMZL, 2%; splenic marginal zone lymphoma, SMZL, 0.9%), mantle cell lymphoma (MCL, 7%), lymphoplasmacytic lymphoma/Waldenstrom macroglobulinemia (WM, 1.4%), and Burkitt lymphoma (0.8%) (Table 22.1) [3].

In addition, plasma cell myeloma is another relatively common mature B-cell neoplasm that results from malignant transformation of plasma cells (derived from B-lymphocytes specifically for antibody production) into myeloma cells (Table 22.1). Myeloma cells then overproduce monoclonal (M) protein (or paraprotein) that lacks anti-infective activity, clogs the bone marrow, breaks down functional immunoglobulins, and damages the kidneys. Plasma cell myeloma causing multiple lesions/tumors is referred to as multiple myeloma (or Kahler's disease), that causing a single lesion/tumor in the bone is referred to as solitary plasmacytoma of bone, and that appearing outside of the bone (e.g., soft tissue) is referred to as extraosseous plasmacytoma. Furthermore, plasma cell myeloma that produces excess M protein, but does not induce tumor/lesion/symptoms, nor meet other criteria for a myeloma diagnosis, is referred to as monoclonal gammopathy of undetermined significance (MGUS) [4,5].

22.2 Biology

The lymphoreticular system is composed of lymph, vessels that transport lymph, and organs that contain lymphoid tissue (e.g., lymph nodes, spleen, and thymus).

The thymus is a flask-shaped organ consisting of two identical lobes (5 cm by 4 cm in diameter, 15–50 g in weight) located in the anterior superior mediastinum, in front of the heart, and behind the sternum. Each lobe of

Table 22.1 Characteristics of Common Mature B-Cell Neoplasms

Type	Biological Features	Affected Organs/ Tissues	Diagnostic Features
Diffuse large B-cell lymphoma (DLBCL)	Fast-growing, aggressive lymphoma	Lymph nodes or outside of the lymphatic system	Sheets of large atypical lymphoid cells (larger than normal B cells); typical immunophenotype: CD20+, CD45+, CD3–
Follicular lymphoma (FL)	Usually slow growing (indolent), transformation to DLBCL in 20%–30% of cases	Germinal center or follicle of lymph node and bone marrow involvement	Closely packed follicles containing small cleaved cells without nucleoli (centrocytes) and larger noncleaved cells with moderate cytoplasm, open chromatin, and multiple nucleoli (centroblasts); typical immunophenotype: CD10+, BCL2+, CD5-, CD20+, BCL6+; rare cases CD10– or BCL2–; t(14;18)(q32;q21)
Chronic lymphocytic leukemia (CLL)/ small lymphocytic lymphoma (SLL)	CLL and SLL represent the early and late stages of the same disease and differ by the tissues affected (CLL in blood and bone marrow; SLL in lymph node)	Blood, bone marrow, and lymph node	Small, mature lymphocytes admixed with larger atypical cells, cleaved cells, or prolymphocytes; absolute B-cell count of $\geq 5 \times 10^9$/L; typical immunophenotype: CD5+, CD23+, CD43+/–, CD10–, CD19+, cyclin D1– ; 17p deletion, del 11q, del 13q, and trisomy 12
Marginal zone lymphoma (MZL)	Three basic subtypes: (i) extranodal or mucosa-associated lymphoid tissue (MALT), (ii) nodal (NMZL), and (iii) splenic (SMZL)	Marginal zone of secondary lymphoid follicles in spleen and lymph nodes as well as MALT	Small to medium marginal zone B cells with irregular nuclei, dispersed chromatin, and inconspicuous nucleoli; typical immunophenotype: CD19+, CD20+, CD45+, CD11c (SMZL), CD5–, CD10-, cyclin D1–; clonal rearrangements of IgH and light chains; trisomy 18 and 3; 1q21 or 1q34 alterations

(Continued)

Table 22.1 *(Continued)* Characteristics of Common Mature B-Cell Neoplasms

Type	Biological Features	Affected Organs/ Tissues	Diagnostic Features
Mantle cell lymphoma (MCL)	Initially indolent, later becoming aggressive	Outer edge (mantle zone) of B-cells in lymph node, one or more organs (e.g., gastrointestinal [GI]), and bone marrow	Tumor cells with no nucleoli; no large cells, no proliferation centers; typical immunophenotype: CD5+, CD23-, cyclin D1+; t(11;14)(q13;q32); minimally mutated IGHV and mostly $SOX11^{+}$; or mutated IGHV and mostly $SOX11^{-}$
Lymphoplasmacytic lymphoma/ Waldenstrom macroglobulinemia (WM)	Indolent; high level of serum immunoglobulin M (IgM); lymphoplasmacytic cell infiltration of bone marrow	Bone marrow, spleen, and lymph nodes	Small B cells with plasmacytic differentiation; typical immunophenotype: surface IgM+, CD19+, CD20+, CD22+, CD25+, CD27+, FMC7+, CD5 variable, CD10–, CD23–, CD103–, CD108–; t(9;14)(p13;q32).
Burkitt lymphoma	Aggressive B-cell lymphoma; may be sporadic, endemic, and immunodeficiency-related	Jaw, central nervous system, bone marrow, bowel, kidneys, ovaries, or other organs.	Monotonous, round, medium-sized cellular morphology; typical immunophenotype: CD10+, BCL6+, BCL2–; c-MYC rearrangement
Plasma cell myeloma	Clonal expansion of malignant plasma cells (i.e., myeloma cells) leads to overproduction of monoclonal (M) protein	Bone, and soft tissue	Multiple myeloma is characterized by the presence of clonal bone marrow plasma cells ≥10% or biopsy-proven bony or extramedullary plasmacytoma, and any one or more of the following events: serum calcium >2.75 mmol/L (>11 mg/dL); serum creatinine >177 μmol/L (>2 mg/dL); hemoglobin <100 g/L; ≥1 osteolytic lesions on CT; any one or more of the following biomarkers of malignancy (clonal bone marrow plasma cell percentage ≥60%; involved:uninvolved serum free light chain ratio ≥100; >1 focal lesions on MRI)

the thymus is divided into a central medulla and a peripheral cortex surrounded by an outer capsule. The cortex contains abundant immature T-cells that migrate to the medulla for maturation. In addition, the thymus consists of small populations of neutrophils, eosinophils, macrophages, and B-lymphocytes.

The spleen is an organ (7 to 14 cm in length, 150 to 200 g in weight) located between the fundus of the stomach and the diaphragm. Composed of red pulp (75% of splenic volume and half of the body's monocytes) and white pulp separated by a marginal zone, the spleen produces blood cells during fetal life (while the bone marrow is solely responsible for hematopoiesis after birth).

The lymph node comprises lymphoid follicles with an outer portion (the cortex, containing immature T cells or thymocytes) and an inner portion (the medulla). The human body possesses about 500 to 600 lymph nodes, commonly located in the mediastinum in the chest, neck, pelvis, axilla, inguinal region, and in association with the blood vessels of the intestines. As an organized collection of lymphoid tissue, the lymph node facilitates lymph passing on its way back to the blood.

Lymphoid tissue consists of connective tissue formed of reticular fibers. Regions of the lymphoid tissue are densely packed with lymphocytes (so called lymphoid follicles). Structurally, lymphoid tissue may form well-organized lymph nodes or loosely organized lymphoid follicles (known as MALT).

The lymphatic vessels include the tubular vessels of the lymph capillaries, the larger collecting vessels (the right lymphatic duct), and the thoracic duct (the left lymphatic duct); and these allow lymph to pass among different parts of the body.

Arising from the lymphatic system, a lymphoid malignancy is called "leukemia" when it is in the blood or marrow or "lymphoma" when it is in lymphatic tissue (e.g., the lymph nodes, spleen, thymus, lymph tissue in the stomach, intestines, and the skin).

Lymphoma is generally classified as either Hodgkin lymphoma or NHL. Hodgkin lymphoma is characterized by the presence of Reed–Sternberg cells in the lymph nodes, association with past Epstein–Barr virus infection, painless "rubbery" lymphadenopathy, occurrence in younger patients, and a favorable prognosis. NHL is characterized by the absence of Reed–Sternberg cells, increased proliferation of B-cells or T-cells, occurrence in older patients, and an unfavorable prognosis.

NHL consists of at least 65 closely related types of lymphomas, which are separated into mature B-cell lymphomas (developing from abnormal B-lymphocytes) and mature T and NK lymphomas (developing from abnormal T-lymphocytes and natural killer cells) (see Chapter 24) [2].

Clonal expansion of malignant plasma cells (myeloma cells) is a multistepped process: (i) binding of specific adhesion molecules on the surface of myeloma cells to stromal cells in the bone marrow induces the production of interleukin-6 (IL-6) by stromal cells, and IL-1-β, tumor necrosis factor-β (TNF-β) and vascular endothelial growth factor (VEGF) by myeloma cells; (ii) IL-6 enhances the replication of myeloma cells; (iii) IL-1- β and TNF-β activate the development of osteoclasts, which break down bone; (iv) VEGF promotes angiogenesis and creates new blood vessels for myeloma development; (v) mature myeloma cells fail to activate the immune system, resulting in a growing number of myeloma cells [4].

22.3 Epidemiology

NHL is the most common cancer of the lymphatic system, and occurs eight times more frequently than Hodgkin lymphoma. NHL generally affects people older than 50 (the average age at diagnosis is 60 to 65), although primary mediastinal (thymic) large B-cell lymphoma is found mostly in young adults (25–40 years of age). Burkitt lymphoma comprises 1% of adult NHL (average age is 30–50 years) but up to 30% of childhood NHL (average age is 5–10 years). DLBCL has an annual incidence of seven to eight cases per 100,000, while CLL/SLL has an annual incidence of 4.7 per 100,000. Plasma cell myeloma (or multiple myeloma) is the third most common blood cancer (after lymphoma and leukemia) in North America, accounting for about 1.4% of the estimated new cancer cases in 2014. Occurring mainly in people aged 60 or over (75%), and rarely in people under 40, it demonstrates a male predilection [5].

22.4 Pathogenesis

Risk factors for NHL include immune system impairment (autoimmune disease); exposure to pesticides, herbicides, fertilizers, and solvents; infection with viruses (EBV, human T-lymphotropic virus type 1, human immunodeficiency virus [HIV], and hepatitis C) and bacteria (*Helicobacter pylori*); and a family history of NHL. In particular, Burkitt lymphoma is linked to EBV in nearly 100% of African cases and about 30% of Western cases. Furthermore, it is associated with HIV in 40% of cases.

Cytogenetic deletion of 13q-14, del 17p13 (p53 locus), t(4;14)(p16;q32), t(11;14)(q13;q32), t(14;16)(q32;q23), amp1q21 and hyperdiploidy as well as various gene mutations (e.g., *KRAS, NRAS, BRAF, FAM46C, TP53*, and *DIS3*) have been implicated in the tumorigenesis of plasma cell myeloma [6].

Progressing from the asymptomatic, precursor pathogenic state of MGUS, and then asymptomatic (smoldering) multiple myeloma, multiple myeloma undergoes various pathological changes that are highlighted by enhanced osteoclast activity causing bone damages (bone pain and fractures), increased loss of calcium in serum and urine (hypercalcemia), reduced levels of red and white blood cells in the blood (anemia and infection vulnerability), and kidney damage (increased or decreased urination) [4].

22.5 Clinical features

Patients with mature B-cell neoplasms often display painless swelling of lymph nodes (in the neck, underarm, stomach, or groin), persistent chills/fever, night sweats, unexplained weight loss, lack of energy, and itching. Depending on the organs/tissues involved, other symptoms may include dizziness; headaches; breathlessness; cough; difficulty swallowing (dysphagia); numbness or paralysis in the face, arm, or legs; diarrhea; patches of inflamed skin, and so on. The most notable symptoms of multiple myeloma are bone pain or bone fractures, increased vulnerability to infections, and increased or decreased urination.

22.6 Diagnosis

Diagnosis of mature B-cell neoplasms requires clinical, laboratory (e.g., complete blood count, platelets, lactate dehydrogenase, comprehensive metabolic panel, and uric acid, electrophoresis for serum/urine M protein, immunoelectrophoresis for immunoglobulins, and Freelite™ serum free light chain assay), radiologic (e.g., positron emission tomography and computed tomography [PET-CT]), and histopathologic assessments. As clinical presentations of mature B-cell neoplasms are mostly nonspecific and variable, microscopic examination and immunohistochemical staining of the lymph node and/or bone marrow is vital for confirmation (Table 22.1).

Among plasma cell myeloma variants, multiple myeloma displays ≥3 g/dL M protein in serum and/or urine, and >30% plasma cells in bone marrow, along with anemia, renal failure, hypercalcemia and osteolytic lesions (Table 22.1); MGUS contains <3 g/dL M protein in serum, and <10% plasma cells in bone marrow, without anemia, renal failure, hypercalcemia and osteolytic lesions;

and asymptomatic (smoldering) multiple myeloma has ≥3 g/dL M protein in serum and/or ≥10% plasma cells in bone marrow, with slight anemia, but without renal failure, hypercalcemia and osteolytic lesions.

Mature B-cell neoplasms are staged according to Lugano classification or other systems [7]. Plasma cell myeloma is defined as stage I (ß2-M <3.5 mg/dL and albumin =3.5 g/dL), stage II (neither stage I nor stage III; albumin <3.5 g/dL and ß2-M <3.5 mg/dL; albumin <3.5 g/dL; or ß2-M 3.5–<5.5 mg/dL), or stage III (ß2-M ≥5.5 mg/L) by using the International Staging System (ISS), which is based on the blood levels of beta 2-microglobulin (ß2-M) and albumin. While ß2-M reveals the extent of disease and kidney function, albumin is indicative of overall general health [4,5].

22.7 Treatment

Treatment options for mature B-cell neoplasms include chemotherapy, radiation therapy, biologic therapy, bone marrow or stem cell transplantation, and surgery (under special circumstances, e.g., splenectomy for WM and primarily to obtain a biopsy for diagnostic purposes) [8,9].

Patients with indolent (slow-growing) forms of mature B-cell neoplasms may not require treatment (e.g., CLL) until signs of progression start to cause problems (e.g., SLL). Treatment may work initially but become ineffective next time. Therefore, new or experimental treatment options need to be considered.

Patients with fast-growing (aggressive) lymphomas require prompt treatment. DLBCL is fatal if left untreated and is curable in about 70% of cases with timely and appropriate treatment (e.g., R-CHOP, which consists of rituximab plus cyclophosphamide, doxorubicin, vincristine, and prednisolone).

Current treatment options for plasma cell myeloma (particularly multiple myeloma) include drug therapies (immunomodulatory drugs, proteasome inhibitors, chemotherapy, histone deacetylase inhibitor, and steroids), stem cell transplants (allogeneic, autologous), biphosphonates, radiotherapy, surgery, and alternative therapies [4,5,10]. For MGUS, watchful waiting is appropriate. For solitary plasmacytoma of bone, radiotherapy is preferred. For extraosseous plasmacytoma, surgery followed by radiotherapy is necessary [5].

22.8 Prognosis

The overall 5-year survival rate for mature B-cell neoplasms is about 63%. Although about 70% of DLBCL cases are curable, patients with early-stage

DLBCL have a better prognosis that those with advanced-stage DLBCL. FL had a mean survival of 8–10 years before the advent of rituximab, which has contributed to an increased survival of FL patients. CLL has a 5-year overall survival of 82%. Patients with MALT lymphoma (with a 5-year survival rate of 88.7%) have a better prognosis than those with SMLZ and NMZL (which have 5-year survival rates of 79.7% and 76.5%, respectively). MCL typically progresses after chemotherapy and has a median survival time of 3 years and a 10-year survival rate of 5%–10%. WM has a median survival time of 5 to 11 years.

Multiple myeloma is a highly treatable but rarely curable disease, with a median survival of 43 and 83 months for stages II and III, respectively. Favorable prognostic factors for multiple myeloma include ß2-M <3 mg/mL, albumin ≥3.5 g/dL, lactate dehydrogenase (LDH) 100–190 U/L (age ≤60 y) or 110–210 U/L (age >60 y), free light chain ratio 0.03–32, and absence of chromosomal abnormalities. Solitary plasmacytoma of bone or extramedullary plasmacytoma are potentially curable. As a benign form of plasma cell myeloma, MGUS requires no treatment. However, MGUS has a 20%–25% lifetime chance of progressing to multiple myeloma or another malignant plasma cell disease (lymphoma or amyloidosis). Additionally, MGUS may be associated with osteoporosis, venous thrombosis and peripheral neuropathy [5].

References

1. Swerdlow SH, Campo E, Harris NL, et al. (eds). *WHO classification of tumours of haematopoietic and lymphoid tissues.* International Agency for Research on Cancer; World Health Organization. Lyon, France: IARC, 2008.
2. Swerdlow SH, Campo E, Pileri SA, et al. The 2016 revision of the World Health Organization classification of lymphoid neoplasms. *Blood.* 2016;127(20):2375.
3. Zelenetz AD, Gordon LI, Wierda WG, et al. Non-Hodgkin's lymphomas, version 4.2014. *J Natl Compr Canc Netw.* 2014;12(9):1282–303.
4. Fairfield H, Falank C, Avery L, Reagan MR. Multiple myeloma in the marrow: pathogenesis and treatments. *Ann N Y Acad Sci.* 2016;1364:32–51.
5. PDQ Adult Treatment Editorial Board. Plasma Cell Neoplasms (Including Multiple Myeloma) Treatment (PDQ®): Health Professional Version. PDQ Cancer Information Summaries [Internet]. Bethesda (MD): National Cancer Institute (US); 2002–2017.
6. Hanbali A, Hassanein M, Rasheed W, Aljurf M, Alsharif F. The evolution of prognostic factors in multiple myeloma. *Adv Hematol.* 2017;2017:4812637.

7. Cheson BD, Fisher RI, Barrington SF, et al. Recommendations for initial evaluation, staging, and response assessment of Hodgkin and non-Hodgkin lymphoma: The Lugano classification. *J Clin Oncol.* 2014;32(27):3059–68.
8. PDQ Adult Treatment Editorial Board. *Adult Non-Hodgkin Lymphoma Treatment (PDQ®): Health Professional Version.* PDQ Cancer Information Summaries [Internet]. Bethesda, MD: National Cancer Institute (US); 2002–2017.
9. PDQ Pediatric Treatment Editorial Board. *Childhood Non-Hodgkin Lymphoma Treatment (PDQ®): Health Professional Version.* PDQ Cancer Information Summaries [Internet]. Bethesda, MD: National Cancer Institute (US); 2002–2016.
10. Raza S, Safyan RA, Rosenbaum E, Bowman AS, Lentzsch S. Optimizing current and emerging therapies in multiple myeloma: a guide for the hematologist. *Ther Adv Hematol.* 2017;8(2):55–70.

23 Hairy Cell Leukemia

Estella Matutes

23.1 Definition

Hairy cell leukemia (HCL) is a mature B-cell lymphoproliferative disorder included as a distinct entity in the World Health Organization (WHO) classification of hemopoietic and lymphoid tumors. It is characterized by the clonal expansion of mature B-lymphocytes with abundant cytoplasm and "hairy" projections involving the peripheral blood (PB), bone marrow (BM), and splenic red pulp. A variant form HCL-variant is considered by the WHO as a provisional entity within the group of unclassifiable splenic small B-cell leukemia/lymphomas, and it is biologically unrelated to HCL [1].

23.2 Biology and pathogenesis

HCL has no distinct chromosomal abnormality. The great majority of cases have somatic hypermutations of the immunoglobulin heavy chain (IGH) gene supporting that the disease arises on a memory B lymphocyte. There is no evidence of specific IGHV (variable), IGHD (diversity), or IGHJ (joining) repertoires or stereotypes in HCL. A major advance in the understanding the molecular pathogenesis of HCL was the discovery of a mutation on the *BRAFV600E* gene as detected by whole exome sequencing in virtually all patients with HCL. The *BRAF* gene is a member of the Raf family that encodes serine/threonine kinases that act in the *MAPK* (mitogen activated protein kinase) pathway. The oncogenic activating mutation of *BRAF* leads to the activation of the *MAPK* signaling pathway as well as the activation of a cascade of kinases. Mutations of *BRAF* have been described in several cancers such as melanoma and thyroid carcinoma but have been rarely documented in hemopoietic malignancies with the exception of Langerhans cell histiocytosis. The finding of *BRAFV600E* mutations is not only relevant to the pathogenesis of HCL, but also has potential therapeutic implications (see Section 23.6). Recently, inactivating mutations on the cell cycle inhibitor *CDKN1b* (p27) concomitantly with *BRAF* mutation in 16% of HCL have been documented. This is the second most frequently mutated gene in HCL. In addition, a small group of HCL that uses the V4-34 IG gene family and lacks mutations on *BRAF* appears to have activating mutations in the *MAP2K1* gene encoding MEK1. Mutations of the *MAP2K1* gene are rare in hemopoietic tumors and so far have only been described in Langerhans cell histiocytosis, HCL-variant, and splenic diffuse red pulp lymphoma (SDRPL) [2–4].

23.3 Epidemiology

HCL is a rare disease with an estimated incidence of around 0.5 per 100,000 per annum. It occurs all around the world, but has a low incidence in Hong Kong and Mexico. In the United States, HCL is observed with a higher frequency among whites compared with blacks or Asians. Several environmental factors including pesticides, herbicides (e.g., the "Agent Orange" used during the Vietnam War), petroleum, and ionizing radiation have been documented to confer a risk for the development of HCL.

HCL affects predominantly middle-aged patients in their 50 s, and is significantly more frequent in males with a male to female ratio of 4:1.

23.4 Clinical features

Patients with HCL may be asymptomatic, and the disease is discovered on a routine checkup. Alternatively, patients manifest symptoms associated with cytopenias such as opportunistic infections, fatigue, bleeding, and/or abdominal discomfort due to the spleen enlargement. Other rare manifestations are cutaneous and bone lytic lesions, gastrointestinal or central nervous system involvement, and/or autoimmune disorders. Peripheral lymphadenopathy is very rare, but around 10%–15% of patients may have abdominal lymphadenopathy. Therefore, CT scan is recommended in the diagnostic and staging workup. The blood counts show variable degrees of anemia, thrombocytopenia, and neutropenia. Macrocytosis is frequent, and monocytopenia is characteristic in active HCL. Most cases have a few circulating lymphocytes with the morphology of hairy cells. Liver function tests may show a raised alkaline phosphatase that often correlates with liver involvement by the disease. In addition to the diagnostic tests on blood and bone marrow, the European Society for Medical Oncology (ESMO) guidelines recommend the following investigations for a staging workup: full blood counts with differential and reticulocytes, liver and renal biochemistry, serum immunoglobulins, B2 microglobulin, direct antiglobulin test (DAT), hepatitis B and C serology, and CT scan of the chest, abdomen, and pelvis [5].

23.5 Diagnosis

In the majority of patients, the diagnosis can be established by morphological examination of blood, immunophenotyping by flow cytometry, and bone marrow trephine biopsy (Figure 23.1). The circulating atypical lymphoid cells are medium in size with a round or kidney-shaped nucleus, loose chromatin, and abundant pale cytoplasm with fine projections. Hairy cells express a variety of B-cell antigens (CD19, CD20, and CD22) and surface immunoglobulin with light chain restriction and bright intensity. The monoclonal antibodies CD11c, CD25, CD103, and CD123 are positive in the majority of cases. Although they are not specific to HCL and may be positive in other splenomegalic B-cell cell disorders, coexpression of the four antigens is almost unique to HCL. In addition, hairy cells have a strong expression of CD200 and CD305. The bone marrow trephine biopsy shows a variable degree of infiltration, ranging from interstitial to diffuse. The neoplastic cells are surrounded by a pale halo due to the abundant cytoplasm giving the typical fried egg appearance. The infiltration is highlighted by immunohistochemistry with the monoclonal antibodies CD20, CD72 (DBA44), CD11c, CD25, annexin A1, and anti-tartrate-resistant acid phosphatase.

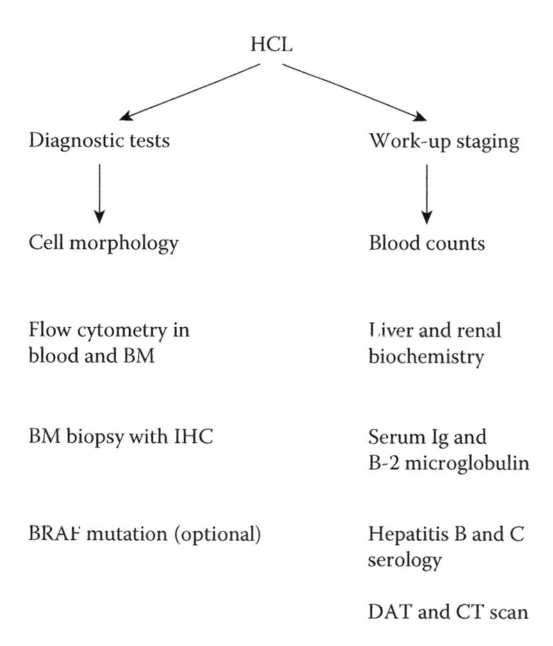

Figure 23.1 Diagnostic approaches for hairy cell leukemia. BM: bone marrow; CT: computerized tomography; DAT: direct antiglobulin test; Ig: immunoglobulins.

Cyclin D1 is usually positive as well as the monoclonal antibody that detects the mutated *BRAF* protein. The latter is useful for detection of minimal residual disease. Cytogenetics are not helpful for diagnosis as there is no evidence for a recurrent chromosomal abnormality to be present in HCL. In cases with atypical features and diagnostic difficulties, detection of *BRAF* mutation can be very valuable in establishing a definitive diagnosis [1,5].

The differential diagnosis arises with other primary splenomegalic disorders such as splenic marginal zone lymphoma, unclassifiable splenic B-cell leukemias/lymphomas, HCL-variant, and—rarely—with primary myelofibrosis and aplastic anemia.

23.6 Treatment

Major advances have been made over the last three decades in the treatment and management of HCL (Figure 23.2). In the minority of patients who are asymptomatic without cytopenias, treatment may be deferred, but a close monitoring of the blood counts every 3 months is recommended. The majority of patients, however, will require treatment at diagnosis due to cytopenias, recurrent infections, symptomatic splenomegaly, and/or systemic symptoms. The two purine nucleoside analogues, 2-chlorodeoxyadenosine (CDA) and 2-deoxycoformycin (DCF), are the drugs recommended for initial treatment. CDA is administered either as a continuous intravenous infusion at a dose of 0.09 mg/kg over a 5- to 7-day period or subcutaneously at a dose of 0.14 mg/kg/day for 5 days. There are no randomized trials

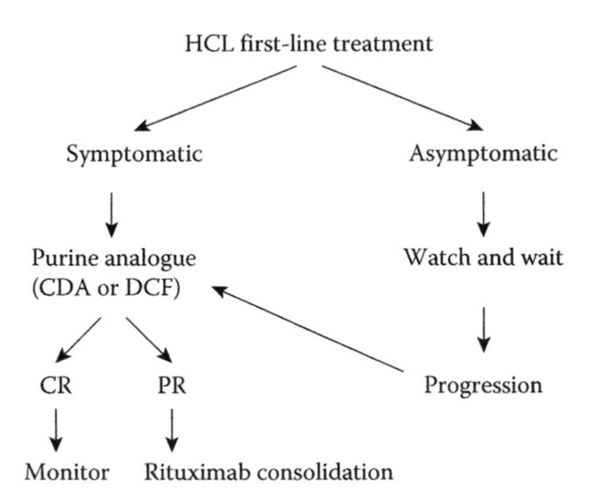

Figure 23.2 Treatment of hairy cell leukemia. CDA: chlorodeoxyadenosine; CR: complete response; DCF: deoxycoformycin; PR: partial response.

comparing the intravenous with the subcutaneous routes, but the efficacy appears to be similar. Responses to CDA are evaluated 3 months after the cycle. The majority of patients achieve a complete response (CR). If a partial response (PR) is obtained, a second cycle with the same dose of CDA is recommended to be given at 3–6 months from the first cycle. The other purine nucleoside analogue, DCF, is given intravenously at a dose of 4 mg/m^2 every 2 weeks. The response is evaluated after normalization of the blood counts 2 weeks after the last infusion and, if a CR is achieved, a further one or two doses are recommended. The overall response rate with CDA or DCF ranges from 85% to 100% with a high CR rate of around 85%. Again, no randomized trials comparing the two purine nucleoside analogues have been performed, but it appears that these analogues induce similar response rates and duration of responses without differences in adverse events. Patients should receive prophylaxis with cotrimoxazole (480 mg/po/daily) and acyclovir 200 mg/po/tds. Granulocyte colony stimulating factor (G-CSF) can be considered in patients with severe neutropenia. Response evaluation includes physical examination, blood counts, bone marrow trephine biopsy with immunohistochemistry, and a CT scan if abdominal lymphadenopathy is present [6–8].

Patients who experience a relapse after 18–24 months from the initial purine nucleoside analogue treatment can be retreated successfully with the same or the alternative purine analogue. The duration of responses is similar to those achieved previously, although the CR rates are lower. Therefore, in this setting, the addition of Rituximab at a dose of 375 mg/m^2 for four to eight doses to the purine analogue is recommended. There are several approaches for the patients who are refractory and/or only achieve transient PR to the purine nucleoside analogues. These include moxetumomab pasudotox, an anti-CD22 recombinant immunotoxin that contains a fragment of a monoclonal antibody (CD22) with a bacterial toxin, Bruton's tyrosine kinase (BTK) inhibitors such as ibrutinib; and agents targeting the mutated *BRAF* gene, such as vemurafenib or dabrafenib. The optimal dosing and duration of treatment with vemurafenib is unknown. Two phase II clinical trials are ongoing in the United States with vemurafenib and ibrutinib in relapsed/refractory HCL patients [6–8].

23.7 Prognosis

Since the introduction of effective treatments three decades ago, the prognosis for HCL is good overall. Patients with HCL have an OS equivalent to the expected age/sex matched general population. At 15 years, the OS is around 85%. In addition, only a minority of deaths are related to HCL due to treatment refractoriness, while the majority are due to other causes such as cancer, vascular disease, and/or older age. Clinical variables that have been reported to influence outcomes include degree of anemia and thrombocytopenia and achievement of only PR and not CR to the purine nucleoside analogue. Among biological variables, usage of V4-34 family and mutations of the tumor suppressor gene *TP53*, which are seen in a minority of patients, predict for an adverse outcome. These latter patients with classical HCL have a poor response to the purine nucleoside analogues similar to the HCL-variant patients [5].

References

1. Matutes E, Martínez-Trillos A, Campo E. Hairy cell leukaemia-variant: Disease features and treatment. *Best Pract Res Clin Haematol.* 2015;28(4):253–63.
2. Tiacci E, Trifonov V, Schiavoni G, et al. BRAF mutations in hairy-cell leukemia. *N Engl J Med.* 2011;364:2305–15.
3. Waterfall JJ, Arons E, Walker RL, et al. High prevalence of MAP2K1 mutations in variant and IGHV4-34-expressing hairy-cell leukemias. *Nat Genet.* 2014;46:8–10.
4. Dietrich S, Hüllein J, Lee SC, et al. Recurrent CDKN1B (p27) mutations in hairy cell leukemia. *Blood.* 2015;126(8):1005–8.
5. Robak T, Matutes E, Catovsky D, Zinzani PL, Buske C, ESMO Guidelines Committee. Hairy cell leukaemia: ESMO clinical practice guidelines for diagnosis, treatment and follow-up. *Ann Oncol.* 2015;26 Suppl 5:v100–7.
6. Else M, Dearden CE, Catovsky D. Long-term follow-up after purine analogue therapy in hairy cell leukaemia. *Best Pract Res Clin Haematol.* 2015;28(4):217–29.
7. Kreitman RJ, Wilson W, Calvo KR, et al. Cladribine with immediate rituximab for the treatment of patients with variant hairy cell leukemia. *Clin Canc Res.* 2013;19:6873–81.
8. Kreitman RJ, Tallman MS, Robak T, et al. Phase I trial of anti-CD22 recombinant immunotoxin moxetumomab pasudotox (CAT-8015 or HA22) in patients with hairy cell leukemia. *J Clin Oncol.* 2012;30:1822–8.

24 Mature T and NK Neoplasms

24.1 Definition

Forming one of the five groups within mature lymphoid, histiocytic, and dendritic neoplasms (see Chapter 22), mature T and natural killer (NK) neoplasms represent about 15% of non-Hodgkin lymphoma (NHL) and consist of at least 20 types, including T-cell prolymphocytic leukemia, T-cell large granular lymphocytic leukemia—chronic lymphoproliferative disorder of NK cells, aggressive NK-cell leukemia, systemic Epstein–Barr virus (EBV)+ T-cell lymphoma of childhood, hydroa vacciniforme–like lymphoproliferative disorder, adult T-cell leukemia/lymphoma, extranodal NK-/T-cell lymphoma nasal type, enteropathy-associated T-cell lymphoma, monomorphic epitheliotropic intestinal T-cell lymphoma—indolent T-cell lymphoproliferative disorder of the gastrointestinal (GI) tract, hepatosplenic T-cell lymphoma, subcutaneous panniculitis-like T-cell lymphoma, mycosis fungoides (MF), Sézary syndrome, primary cutaneous CD30+ T-cell lymphoproliferative disorders, lymphomatoid papulosis, primary cutaneous anaplastic large cell lymphoma (primary cutaneous CD8+ aggressive epidermotropic cytotoxic T-cell lymphoma, primary cutaneous acral CD8+ T-cell lymphoma, and primary cutaneous CD4+ small/medium T-cell lymphoproliferative disorder), peripheral T-cell lymphoma not otherwise specified (NOS), angioimmunoblastic T-cell lymphoma (follicular T-cell lymphoma, nodal peripheral T-cell lymphoma with TFH phenotype), anaplastic large-cell lymphoma, ALK+, and anaplastic large-cell lymphoma, ALK– (breast implant–associated anaplastic large-cell lymphoma) [1,2].

The most common mature T and NK neoplasms are cutaneous T-cell lymphomas (cutaneous T-cell lymphomas [CTCL]; including MF, Sezary syndrome; representing about 5% of NHL), extranodal natural killer/T-cell lymphoma nasal type (ENKTCL, 5%), angioimmunoblastic T-cell lymphoma (AITL, 2%), adult T-cell leukemia/lymphoma (ATLL; including smoldering, chronic, acute, and lymphoma subtypes), enteropathy-associated intestinal T-cell lymphoma (EATL, Types 1 and 2), anaplastic large cell lymphoma (ALCL; primary cutaneous, systemic, breast implant–associated; 2%), and peripheral T-cell lymphoma not otherwise specified (PTCL NOS) (Table 24.1).

Table 24.1 Characteristics of Common Mature T and NK Neoplasms

Type	Biological Features	Affected Organs/Tissues	Diagnostic Criteria
Cutaneous T-cell lymphomas (CTCL)	Mycosis fungoides (MF, most common subtype, indolent); Sézary syndrome (SS, advanced or variant MF)	Skin and blood	**MF**—typical immunophenotype: CD2+, CD3+, CD4+, CD5+, CD8–, CD45RO+, CD20–, and CD30–; CD4– or CD8+, rare double positive or double negative cases; Th2 profile; TCR-αβ+. **SS**—>1,000 Sézary cells per mm^3; Sézary cells (>14 µm in diameter) >20% of circulating lymphocytes; typical immunophenotype: CD3+, CD4+, CD7–, and CD8–
Extranodal natural killer (NK)/T-cell lymphoma, nasal type (ENKTCL)	Highly aggressive; associated with Epstein–Barr virus (EBV)	Nasal cavity, paranasal sinuses, and nasopharynx	Medium-sized or mixture of small and large neoplastic lymphoid cells; infiltration of blood vessel walls; prominent necrosis; typical immunophenotype: (CD)3ε+, CD56+, EBV+; CD5–, CD20–
Angioimmunoblastic T-cell lymphoma (AITL)	Often aggressive; associated with EBV (but neoplastic T cells EBV-); occurring after penicillin use (27%)	Lymph node, bone marrow, spleen, liver, skin, and pleuropulmonary	Perinodal infiltration, vascular proliferation, follicular dendritic cell proliferation, bone marrow involvement, reactive lymphocytes and immunoblasts in blood; typical immunophenotype: CD10+/–, CXCL13+, BCL–6+/–, PD1+, CD4+, or mixed CD4/8, EBV+CD20+ B blasts

(Continued)

Table 24.1 *(Continued)* Characteristics of Common Mature T and NK Neoplasms

Type	Biological Features	Affected Organs/Tissues	Diagnostic Criteria
Adult T-cell leukemia/lymphoma (ATLL)	Aggressive (acute and lymphomatous), less aggressive (chronic and smoldering); associated with HTLV	Blood (leukemia), lymph node (lymphoma), skin, or multiple areas of the body	*Acute:* leukemic picture, organomegaly, high LDH and often hypercalcemia. *Chronic:* lymphocytosis >4 × 10^9/L with ATLL cells; skin, lung, liver, or node involvement. *Smouldering:* skin and/or lung infiltrates; normal lymphocyte count (1 %–5% ATLL cells). *Lymphoma:* organomegaly; <1% circulating leukemic cells; high LDH and possible hypercalcemia. Typical immunophenotype: CD4+ CD25+, CD7–, CD30–/+, CD15–/+
Enteropathy-associated intestinal T-cell lymphoma (EATL)	Type I (80%) and type II (20%); strong association with celiac disease	Frequently jejunum or ileum, also duodenum, stomach, colon, liver, gallbladder, spleen, and lymph node	*Type I:* medium to large tumor cells, round or angulated vesicular nuclei, prominent nucleoli, pale cytoplasm; typical immunophenotype: CD3+, CD5–, CD7+, CD8+/–, CD4–, CD30+; TCRγ. *Type II:* multi foci of small, round, uniform cells, dark nuclei, pale-rimmed cytoplasm; typical immunophenotype: CD3+, CD8+, CD56+, CD4–; TCR-β; 8q24 (MYC)
Anaplastic large cell lymphoma (ALCL)	Three subtypes (primary cutaneous, systemic, breast implant-associated)	Skin, lymph node, or other organs	Anaplastic cytology (large tumor cells, abundant cytoplasm, prominent nucleoli, pleomorphic nucleus; infiltration of lymph node); typical immunophenotype: CD30+, CD25+, CD4+/–, CD3–/+, CD43+; ALK+ or t(2;5) in 60%–70% systemic cases

24.2 Biology

Although the spleen is involved in the production of blood cells during fetal life, the bone marrow is solely responsible for this function after birth. Located inside the bone marrow, hematopoietic stem cells give rise to myeloid and lymphoid stem cells. Myeloid stem cells further differentiate into erythroblast, myelobalst, monoblast, and megakaryoblast, which in turn produce erythrocyte, eosinophil, basophil, neutrophil, mast cell, and monocyte, as well as platelets. On the contrary, lymphoid stem cells evolve into B-, and T-lymphoblasts and NK cell precursor, which then become B-, and T-lymphocytes and NK cells (see Figure 13.1).

Lymphoid malignancies are derived from the cells and tissues within the lymphatic system. Those involving the blood or bone marrow are called "leukemia," and those involving lymphatic tissue (e.g., lymph nodes, spleen, thymus, and lymph tissue in the stomach, intestines, and skin) are called "lymphoma."

Lymphoma is generally classified as either Hodgkin lymphoma or NHL. Consisting of at least 65 closely related types, NHL is separated into mature B-cell lymphomas (developing from abnormal B-lymphocytes; 45 types; see Chapter 22) and mature T and NK lymphomas (developing from abnormal T-lymphocytes and NK cells, 20 types) [2].

Among most common T and NK lymphomas, CTCL are the second most frequent extranodal NHL (after gastrointestinal lymphomas). Compared to nodal NHL, which is largely of the B type, extranodal NHL is mostly of the T type. The most common subtype of CTCL is MF, which usually progresses from localized skin lesions to systemic disease. Sézary syndrome is regarded as an advanced variant of MF and is characterized by generalized erythroderma and atypical circulating lymphocytes (Sézary or Lutzner cells) in the peripheral blood.

Extranodal NK-/T-cell lymphoma (ENKTCL), nasal type is thought to originate from mature NK cells as the vast majority of tumor cells express an NK cell phenotype (although a small number of cells express a cytotoxic T-cell phenotype). Typically affecting the nasal cavity, ENKTCL is characterized by progressive midline facial destruction.

Angioimmunoblastic T-cell lymphoma (AITL, also known as angioimmunoblastic lymphadenopathy with dysproteinemia, or AILD) is characterized by a polymorphous lymph node infiltrate with a marked increase in follicular dendritic cells and high endothelial venules and systemic involvement. It can be subdivided into three subtypes: (i) AITL with no evidence of clonal lymphoid proliferation, (ii) AITL-type dysplasia with inconsistent clonality, and (iii) AILD-type lymphoma with strong evidence of clonality.

Adult T-cell leukemia/lymphoma (ATLL) is a mature T-lymphoid malignancy with an etiological link to human T-cell lymphotropic virus 1 (HTLV-1). It includes both aggressive subtypes (acute 65% and lymphoma) and indolent subtypes (chronic and smoldering) and can appear in the blood (leukemia), lymph nodes (lymphoma), skin, or multiple areas of the body.

Enteropathy associated intestinal T-cell lymphoma (EATL) usually presents as a tumor composed of large lymphoid cells, often with an inflammatory background in the jejunum or ileum. Based on histology, immunophenotype, and relationship to celiac disease (CD), EATL is divided into two subtypes (Type I, 80%, and Type II, 20%).

ALCL is identified by its anaplastic cytology and constant membrane expression of the CD30 antigen (an activation marker for B or T cells). It can be distinguished into three subtypes: primary cutaneous, systemic (ALK+ and ALK–), and breast implant–associated.

PTCL NOS covers a morphologically and immunophenotypically heterogeneous group of nodal T-cell lymphomas that do not meet the diagnostic criteria for one of the other World Health Organization (WHO)-defined T-cell lymphomas.

24.3 Epidemiology

CTCL mainly affects people aged 55–60 years. Within CTCL, MF has a male to female ratio of 1.6–2:1 and an annual incidence of 0.36 per 100,000 cases in the United States.

ENKTCL affects predominantly middle-aged males (with a median age of 44 years, male to female ratio of 3:1). Linked to EBV, ENKTCL is more common in Asia (particularly China, Korea, Hong Kong, and Japan) and in South America, where it accounts for 7%–10% of all NHL, but ENKTCL only accounts for 1% of NHL in other countries. The annual incidence of ENKTCL in the United States is 0.036 per 100,000 cases.

AITL (AILD) represents 2% of NHL and 15%–20% of PTCL and usually occurs in people aged 40–90 years without gender preference.

ATLL almost exclusively affects adults and rarely children in HTLV-1-endemic areas (e.g., southwestern Japan [those around 60 years old], the Caribbean Islands, Central and South America, West Africa, and Middle East [those around 40 years old]). The estimated lifetime risk of developing ATLL in HTLV-1 carriers is 6%–7% for men and 2%–3% for women in Japan.

EATL tends to occur in the 40–70 age group (with a median age of 60 years, male to female ratio of 1.2–1.4:1) and has an annual incidence of 1 per million worldwide. EATL type I accounts for two in three cases and is more common in Europe. EATL type II accounts for one in three cases and is more common in Asia. However, Types I and II are equally common in North America.

ALCL comprises about 3% of adult NHL but up to 30% of childhood NHL. Patients with primary cutaneous-ALCL and ALK– systemic ALCL are generally older (with a median age of 61 years) than patients with the ALK+ systemic ALCL (with a median age of 24 years). Patients with breast implant–associated ALCL are diagnosed at 53 years (the range is 25–91 years).

PTCL NOS is the most common nodal T-cell lymphoma (accounting for 30% of such cases in Western countries). It usually affects adults and very rarely children and has a male to female ratio of 2:1.

24.4 Pathogenesis

Risk factors for NHL range from immune system impairment (autoimmune disorders such as CD, dermatitis herpetiformis, and hyposplenism in EATL); exposure to pesticides, herbicides, fertilizers, and solvents; infection with viruses (EBV, HTLV-1, and human immunodeficiency virus [HIV]); and a family history of NHL, to genetic mutations (e.g., t[2;5] in ALCL).

24.5 Clinical features

Patients with mature T and NK neoplasms often show painless swelling of lymph nodes (in the neck, underarm, stomach, or groin), persistent chills/fever, night sweats, unexplained weight loss, lack of energy, and itching. Other symptoms may include dizziness; headaches; breathlessness; cough; difficulty swallowing (dysphagia, e.g., ENKTCL); numbness or paralysis in the face, arm, or legs; diarrhea (e.g., EATL); patches of inflamed skin; and pruritus (e.g., MF, Sézary syndrome, and AILD).

24.6 Diagnosis

Diagnosis of mature T and NK neoplasms relies on clinical, laboratory (e.g., complete blood count, platelets, lactate dehydrogenase, comprehensive metabolic panel, and uric acid), radiologic (e.g., positron emission tomography and computed tomography [PET-CT]), and pathologic examinations. Due to the fact that clinical presentations of mature T and NK neoplasms are nonspecific and highly variable, microscopic, immunohistochemical, and molecular examination of lymph nodes and/or bone marrow is essential for precise determination (Table 24.1) [3].

24.7 Treatment

Treatment of mature T and NK neoplasms involves chemotherapy, radiation therapy, biologic therapy, bone marrow or stem cell transplantation, and surgery [4]. For CTCL, treatment is directed either at the skin (e.g., ultraviolet light, topical steroids, topical chemotherapies, topical retinoids, and electron beam radiation therapy) or the entire body (e.g., oral retinoids, photopheresis, fusion proteins, interferon, and systemic chemotherapy). For AITL, treatment includes a combination of a steroid and a multiagent chemotherapy regimen (e.g., cyclophosphamide/hydroxydaunomycin/vincristine/prednisone [CHOP]). For ATLL, watchful waiting may be appropriate for patients with mild or no symptoms, while skin-directed therapies may be prescribed for those with skin lesions. For EATL, CHOP is commonly administered. Although radiation therapy or surgical excision may be utilized for solitary or localized primary cutaneous ALCL, CHOP represents the most common therapy for systemic ALCL [4].

24.8 Prognosis

While MF has a 5-year survival rate of 80%–100%, Sezary syndrome and AITL have a median survival of 32 and 24 months, respectively. ATLL demonstrates a median survival of less than 12 months in the acute and lymphoma subtypes and a 4-year survival rate of 26.9% and 62% in chronic and smouldering subtypes, respectively. Cutaneous CD30+ ALCL has a 5-year survival rate of 90% compared to 15% in primary cutaneous CD30− ALCL.

References

1. Swerdlow SH, Campo E, Harris NL, et al. (eds). *WHO classification of tumours of haematopoietic and lymphoid tissues*. International Agency for Research on Cancer; World Health Organization. Lyon, France: IARC, 2008.
2. Swerdlow SH, Campo E, Pileri SA, et al. The 2016 revision of the World Health Organization classification of lymphoid neoplasms. *Blood*. 2016;127(20):2375.

3. Cheson BD, Fisher RI, Barrington SF, et al. Recommendations for initial evaluation, staging, and response assessment of Hodgkin and non-Hodgkin lymphoma: The Lugano classification. *J Clin Oncol.* 2014;32(27):3059–68.
4. PDQ Adult Treatment Editorial Board. *Adult Non-Hodgkin Lymphoma Treatment (PDQ®): Health Professional Version.* PDQ Cancer Information Summaries [Internet]. Bethesda (MD): National Cancer Institute (US); 2002. 2017 Jan 26.

25
Hodgkin Lymphoma

25.1 Definition

Classified under the mature lymphoid, histiocytic, and dendritic neoplasms category (which also includes mature B-cell neoplasms, mature T and NK neoplasms, post-transplant lymphoproliferative disorders [PTLD], and histiocytic and dendritic cell neoplasms, of which mature B-cell neoplasms and mature T and natural killer [NK] neoplasms are collectively referred to as non-Hodgkin lymphoma or NHL), Hodgkin lymphoma (HL) makes up about 15% of lymphoma cases and is divided on the basis of histological and phenotypic features of the tumor cells into (i) nodular lymphocyte predominant HL (NLPHL) and (ii) classical HL (CHL).

NLPHL constitutes 5% of HL cases and is a non-classic, indolent tumor showing total replacement of nodal architecture by expansive vague nodules of small lymphocytes and sparse, large lymphocyte predominant cells (LPC).

CHL accounts for 95% of HL cases and is a monoclonal lymphoid neoplasm composed of bi- and multinucleated Hodgkin Reed–Sternberg cells (HRSC) in a background of non-neoplastic small lymphocytes, eosinophils, neutrophils, histiocytes, plasma cells, fibroblasts, and collagen fibers. CHL can be further differentiated into four subtypes: (i) nodular sclerosis CHL (NSCHL, 70% of CHL), (ii) lymphocyte-rich classical HL (LRCHL, 6%), (iii) mixed cellularity classical HL (MCCHL, 23%), and (iv) lymphocyte-depleted classical HL (LDCHL, 1%) (Table 25.1) [1,2].

25.2 Biology

Arising from the lymphatic system, lymphoma is a malignancy that occurs in lymphatic tissue (e.g., lymph nodes, spleen, thymus, and lymph tissue in the stomach, intestines, and skin), while leukemia is a malignancy that affects the blood or bone marrow.

Lymphoma is commonly classified into HL and NHL. Derived from a germinal center B cell, HL is characterized by the presence of neoplastic cells (bi- or multinucleated HRSC or an occasionally mononuclear RS variant called

Table 25.1 Characteristics of Hodgkin Lymphoma

Type	Biological Features	Affected Organs/Tissues	Diagnostic Features
Nodular lymphocyte predominant Hodgkin lymphoma (NLPHL)	Indolent course; nodule formation; high recurrence; transformation into diffuse large B-cell lymphoma (DLBCL) in 3%–5% of cases	Lymph node	Total replacement of nodal architecture by nodules of small lymphocytes and sparse, large lymphocyte predominant cells (LPC; multilobed or round nucleus, dispersed chromatin, small nucleoli, narrow rim of basophilic cytoplasm); small lymphocytes surrounding LPC (CD4+/CD57+/PD1+); typical LPC immunophenotype: CD15–, CD20+, CD30–, CD45+; t(3;14)(q27;q32) in 20% of cases
Nodular sclerosis classical Hodgkin lymphoma (NSCHL)	Nodule formation; two grades (NS-1/NS-2); other variants (cellular phase/syncytial variant)	Thymus	Presence of sclerosis, lacunar cells (with lobated nuclei, small nucleoli, and wide rim of clear or acidophilic cytoplasm) and nodular pattern (with collagenous bands dividing the lymph node into nodules), focal necrosis; typical Hodgkin Reed–Sternberg cell (HRSC) immunophenotype: CD15+, CD20–, CD30+, CD45–
Lymphocyte-rich classical Hodgkin lymphoma (LRCHL)	Nodular or diffuse growth; low aggressiveness; more frequent late relapses	Lymph node	HRSC throughout nodules or at periphery; admixed histiocytes; typical HRSC immunophenotype: CD15+, CD20–, CD30+, CD45–
Mixed cellularity classical Hodgkin lymphoma (MCCHL)	Diffuse growth; two variants (interfollicular/epithelioid cell-rich); Epstein–Barr virus (EBV) associated	Lymph node	Neoplastic elements (HRSC without lacunar or popcorn variants); surrounded by plasma cells, epithelioid histiocytes, eosinophils, and T-lymphocytes (CD3+/CD57–); typical HRSC immunophenotype: CD15+, CD20–, CD30+, CD45–
Lymphocyte-depleted classical Hodgkin lymphoma (LDCHL)	Aggressive; two subtypes (fibrotic and reticular/ sarcomatous); EBV associated	Lymph node, bone marrow involvement	Numerous large HRSC, scanty small lymphocytes, plasma cells, histiocytes, and granulocytes; focal necrosis; typical HRSC immunophenotype: CD15+, CD20–, CD30+, CD45–

Hodgkin cell with single round or oblong nucleus and large inclusion-like nucleoli) and mononucleated lymphocyte-predominant cells (LPC, formerly lymphocytic and histiocytic cells or popcorn cells) in the lymph nodes of CHL and NLPHL, past Epstein–Barr virus (EPV) infection, painless "rubbery" lymphadenopathy, predominantly younger adults, relatively infrequent occurrence (15% of lymphoma cases), and a favorable prognosis. In contrast, NHL is characterized by the absence of HRSC, increased proliferation of B-cells or T-cells, predominantly older adults, frequent occurrence (85% of lymphoma cases), and an unfavorable prognosis [3].

Representing a clonal population, HRSC (20–60 μm in diameter) is a bi- or multi nucleated giant cell with a bilobed nucleus and two large acidophilic nucleoli (a characteristic owl's eye appearance). Despite its B-cell origin, HRSC has lost much of the B-cell-specific markers, with a typical immunophenotype of CD15+, CD20−, CD30+, and CD45−. HRSC synthesizes TNF-R, produces T helper lymphocytes type 2 (Th2) cytokines and chemokines (e.g., IL-5, IL-7, IL-8, IL-9, CCL-5, CCL-17, CCL-20, and CCL-22), and also expresses a broad range of receptors (e.g., CD30, CD40, IL-7R, IL-9R, IL-13R, TACI, and CCR5). LPC differs from HRSC by its possession of a single, multilobulated, or round nucleus and possesses a typical immunophenotype of CD15−, CD20+, CD30−, and CD45+, suggesting that NLPHL is biologically distinct from CHL [3].

HL typically affects lymph nodes and sometimes other organs (e.g., liver, lung, and bone marrow), although NSCHL can develop in the thymus.

25.3 Epidemiology

HL is responsible for about 1% of all cancers and 6% of childhood cancers, with an annual incidence of about three cases per 100,000 and a propensity to affect younger adults. The male-to-female ratio in children younger than 5 years is 5.3:1, and in children aged 15–19 years, the ratio is 0.8:1.

NLPHL mainly occurs in males in the fourth decade of life. CHL demonstrates a bimodal age distribution, with two peaks in the second and seventh decades of life, and shows a male preponderance (with the exception of NSCHL, which predominates in women and forms a mediastinal mass in 80% of patients).

25.4 Pathogenesis

Risk factors for HL include EBV infection (with EBV DNA detected in HRSC from approximately 75% of MCCHL, 10%–40% of NSCHL, and 40% of CHL), immune deficiency, and a family history of HL (particularly when first-degree relatives are diagnosed before the age of 30).

Molecularly, HRSC contains clonal immunoglobulin variable region heavy chain (IGVH) rearrangements in >98% of HL cases (based on single-cell microdissection instead of whole tissue). The tumor cells with rearranged IGVH show a high frequency of somatic hypermutations. Comparative genomic hybridization reveals recurrent gains of the chromosomal subregions on chromosome arms 2p, 9p, and 12q and high-level amplifications on chromosome bands 4p16, 4q23-q24, and 9p23-p24. Indeed, overexpression of the genes encoding for CD274/PD-L1, CD273/PD-L2, and JAK2 within the 9p24.1 region occurs in both CHL and primary mediastinal large B-cell lymphoma (PMBL). Whole-exome sequencing on purified HRSC shows that the most frequently detected gene mutation in CHL is inactivating B2M.

25.5 Clinical features

The clinical symptoms of HL are largely attributed to the numerous cytokines, chemokines, and products of the tumor necrosis factor receptor (TNF-R) family secreted by HRSC.

About 40% of HL patients manifest nonspecific constitutional symptoms (e.g., fatigue, anorexia, weight loss, pruritus, night sweats, fever, and alcohol-induced nodal pain) of which unexplained fever (temperature above 38.0°C orally), unexplained weight loss (10% of body weight within the 6 months preceding diagnosis), and drenching night sweats are commonly referred to as B symptoms.

Clinical symptoms resulting from direct or indirect effects of nodal or extranodal involvement cells (e.g., NLPHL) include painless lymphadenopathy (commonly involving the supraclavicular or cervical area). However, CHL involving mediastinum affects about 75% of adolescents and young adults and is mostly asymptomatic.

25.6 Diagnosis

Diagnosis of HL requires the identification of characteristic neoplastic cells (e.g., HRSC with at least two nucleoli in two separate nuclear lobes in CHL and LPC with a single, multilobed or round nucleus in NLPHL) within an inflammatory milieu (Table 25.1) [4,5]. Molecularly, clonal Ig gene rearrangement is detectable in isolated HRSC DNA (>98% of cases). However, HRSC is not exclusive of HL, and similar elements may be observed in reactive lesions (e.g., infectious mononucleosis), B- and T-cell lymphomas, carcinomas, melanomas, and sarcomas.

The stages (I, IE, II, IIE, III, III1, III2, and IV) of HL are determined after taking into account the tumor size, extranodal involvement, elevated erythrocyte sedimentation rate, B symptoms, and so on [4,5]. HL considered clinically as *early favorable* is clinical stage I or II without any risk factors; that considered *early unfavorable* is clinical stage I or II with one or more of risk factors (large mediastinal mass [>33% of the thoracic width on the chest X-ray, ≥10 cm on CT scan], extranodal involvement, elevated erythrocyte sedimentation rate [>30 mm/h for B stage, >50 mm/h for A stage], three or more lymph node areas' involvement, and B symptoms); and that considered as *advanced* is advanced-stage HL with risk factors (an albumin level of <4.0 g/dL, a hemoglobin level of <10.5 g/dL, male sex, aged 45 or more, stage IV disease, white blood cell [WBC] count of ≥15,000/mm^3, absolute lymphocytic count of <600/mm^3, or a lymphocyte count that is <8% of the total WBC count).

25.7 Treatment

Up to 80% of HL is curable with multiagent chemotherapy (e.g., ABVD [Adriamycin (doxorubicin), bleomycin, vinblastine, dacarbazine], AVD [adriamycin, vinblastine, dacarbazine], BEACOPP [bleomycin, etoposide, doxorubicin, cyclophosphamide, vincristine, procarbazine, prednisone], and MOPP [mechlorethamine, vincristine, procarbazine, prednisone]) often in combination with radiation therapy (usually 25–30 Gy to clinically uninvolved sites and 35–44 Gy to regions of initial nodal involvement) [6–8].

For relapsed or refractory HL, cytotoxic chemotherapy delivered via specific targets on the cell surface (antibody-drug conjugates [ADC]) to the interior of cancer cells and use of checkpoint inhibitors (to enhance the ability of the patient's immune system) may be explored.

The key adverse effect of MOPP- or BEACOPP-containing regimens is infertility, although ABVD appears to spare long-term testicular and ovarian function. Complications related to radiotherapy include hypothyroidism, cardiac disease, and impairment of pulmonary function.

25.8 Prognosis

NLPHL has a 10-year survival rate of 80%–90%, which may be negatively impacted by bone marrow involvement (associated with increased aggressiveness) and diffuse T-cell rich pattern (associated with higher recurrence). CHL has a 5-year survival rate of 60% (Stage IV) to 90% (Stage I or II) and shows an increased risk of second cancers (breast cancer, solid tumors, acute myeloid leukemia [AML], and myelodysplasia) after chemotherapy and/or radiotherapy treatments. Unfavorable prognostic factors for CHL are an advanced stage of disease, presence of B symptoms, presence of bulky disease, extranodal extension, elevated erythrocyte sedimentation rate, leukocytosis (white blood cell count >11,500/mm^3), anemia (hemoglobin <11.0 g/dL), male gender, and poor response to initial chemotherapy.

References

1. Swerdlow SH, Campo E, Pileri SA, et al. The 2016 revision of the World Health Organization classification of lymphoid neoplasms. *Blood.* 2016;127(20):2375.
2. Küppers R, Engert A, Hansmann ML. Hodgkin lymphoma. *J Clin Invest.* 2012;122(10):3439–47.
3. Agostinelli C, Pileri S. Pathobiology of Hodgkin lymphoma. *Mediterr J Hematol Infect Dis.* 2014;6(1):e2014040.
4. Cheson BD, Fisher RI, Barrington SF, et al. Recommendations for initial evaluation, staging, and response assessment of Hodgkin and non-Hodgkin lymphoma: The Lugano classification. *J Clin Oncol.* 2014;32(27):3059–68.
5. PathologyOutlines.com website. *Hodgkin's lymphoma-general.* http://www.pathologyoutlines.com/topic/lymphomanonBhodgkins.html. Accessed February 1, 2017.
6. PDQ Pediatric Treatment Editorial Board. *Childhood Hodgkin Lymphoma Treatment (PDQ®): Health Professional Version.* PDQ Cancer Information Summaries [Internet]. Bethesda, MD: National Cancer Institute (US); 2002–2017.
7. PDQ Adult Treatment Editorial Board. *Adult Hodgkin Lymphoma Treatment (PDQ®): Health Professional Version.* PDQ Cancer Information Summaries [Internet]. Bethesda, MD: National Cancer Institute (US); 2002–2017.
8. Viviani S, Tabanelli V, Pileri SA. The pathobiology and treatment of Hodgkin lymphoma. Where do we go from Gianni Bonadonna's lesson? *Tumori.* 2017;103(2):101–13.

26 Post-Transplant Lymphoproliferative Disorders (PTLD)

26.1 Definition

Grouped together with mature B-cell neoplasms, mature T and NK neoplasms, Hodgkin lymphoma, and histiocytic and dendritic cell neoplasms under the mature lymphoid, histiocytic, and dendritic neoplasms category, post-transplant lymphoproliferative disorders (PTLD) consist of six types: (i) plasmacytic hyperplasia PTLD (PH-PTLD), (ii) infectious mononucleosis PTLD (IM-PTLD), (iii) florid follicular hyperplasia PTLD (FFH-PTLD), (iv) polymorphic PTLD (P-PTLD), (v) monomorphic PTLD (B- and T-/NK-cell types) (M-PTLD), and (vi) classical Hodgkin lymphoma PTLD (CHL-PTLD) (Table 26.1) [1].

26.2 Biology

PTLD represent a heterogeneous group of diseases that occur as serious complications of solid organ transplantation (SOT) and hematopoietic stem cell transplantation (HSCT). Originating mostly from B-cells (85%) and occasionally from T- and natural killer (NK)-cells (15%) and showing a clear association with Epstein–Barr virus (EBV), PTLD can vary pathologically from benign lymphoproliferations to malignant lymphomas.

SOT-related PTLD evolves largely from the postgerminal center host lymphocytes, suggesting a role of the graft and endogenous EBV reactivation in chronic B-cell stimulation. Interestingly, host-derived PTLD following SOT shows a higher risk for persistent and recurrent disease, while donor-derived PTLD has early onset (the first months after transplantation). PTLD post-HSCT arises most commonly from the donor lymphocytes as a result of graft-versus-host disease and develops during the first 6 months after transplant [2,3].

Early-onset PTLD often involves the lymph nodes or the allograft, and late-onset PTLD is more common extranodally.

Table 26.1 Characteristics of Post-Transplant Lymphoproliferative Disorders (PTLD)

Type	Biological Features	Affected Organs/Tissues	Diagnostic Features
Plasmacytic hyperplasia PTLD (PH-PTLD)	Usually regress; some Epstein–Barr virus (EBV)-negative cases resemble nonspecific lymphoid hyperplasia	Tonsils or lymph nodes	Preservation of nodal or tonsilar architecture; proliferation of polytypic plasma cells, small lymphocytes, and interspersed transformed large lymphocytes
Infectious mononucleosis PTLD (IM-PTLD)	Mostly EBV positive; possible progression to aggressive P-PTLD	Tonsils or lymph nodes	Partial reservation of nodal or tonsilar architecture; paracortical hyperplasia; numerous immunoblasts in a background of small T-cells, plasma cells, and hyperplastic follicles
Florid follicular hyperplasia PTLD (FFH-PTLD)	Indistinguishable from nonspecific FFH in normal host; resembling IM-PTLD; potential to develop more advanced lesions	Tonsils or lymph nodes	Partial reservation of nodal or tonsilar architecture; morphological appearance of FFH with insignificant expansion of interfollicular area and few interfollicular immunoblasts/transformed cells; EBV positivity and mass formation in post-transplant host
Polymorphic PTLD (P-PTLD)	Polyclonal/monoclonal; most cases arising within 1 year after transplantation, being monoclonal and EBV-positive; some regress, other progress	Lymph nodes, gastrointestinal tract (GIT), lung, or allograft	Near complete destruction of nodal architecture, necrosis, and nuclear atypia; proliferation of lymphocytes (with clefted or angulated nuclei), plasma cells, histiocytes, and eosinophils; presence of atypical Reed–Sternberg-like forms (CD20+, CD30+, and CD15–)

(Continued)

Table 26.1 *(Continued)* Characteristics of Post-Transplant Lymphoproliferative Disorders (PTLD)

Type	Biological Features	Affected Organs/Tissues	Diagnostic Features
Monomorphic PTLD (B- and T-/NK-cell types) (M-PTLD)	Monoclonal; B-cell type mostly resembling DLBCL; T-cell type mostly resembling peripheral T-cell lymphoma and often EBV negative; most progress rapidly	Lymph node, extranodal site (including bone marrow)	Complete destruction of nodal architecture by sheets of atypical cells; proliferation of B-, T-, natural killer (NK)-cells, plasma cells; dominant clones of large cells; DLBCL-like B-cell type with pan-B-cell markers; T/NK cell types with pan-T-cell and occasionally NK-cell markers; mostly CD20+; trisomy 9 and 11, 8q24 and 14q32 rearrangements, 1q11-21 breaks
Classical Hodgkin lymphoma PTLD (CHL-PTLD)	Resembling mixed cellularity or lymphocyte-depleted CHL but involving mucosa-associated lymphoid tissues (MALT)	Lymph nodes, extranodal site	Destruction of nodal architecture; presence of Reed–Sternberg cells (CD15+, CD30+, CD45–, occasionally CD15–) in a background of plasma cells, eosinophils, and (epithelioid) histiocytes (in contrast to CD15–, CD20+, CD45+ immunoblasts and transformed large cells in other PTLD types)

26.3 Epidemiology

PTLD occur in 2%–10% of SOT and HSCT recipients (ranging from <1% in renal transplants; 1%–2% in bone marrow, stem cell, pancreas, heart, and liver transplants; 2%–6% in combined heart–lung transplants; 4%–10% in lung transplants; and up to 20% in small intestine transplants). The incidence of PTLD in children is generally higher than that in adults as children often have negative serology and thus increased susceptibility for EBV prior to transplantation. In addition, longer duration of post-transplant immunosuppression, greater combined intensity of immunosuppression, and EBV mismatch (i.e., transplantation of an organ or tissue from an EBV-seropositive donor to an EBV-seronegative recipient) are associated with heightened development of PTLD.

The incidence of PTLD has risen during the last decade—due likely to the availability of more potent immune suppression, older ages of donors and recipients, more frequent haploidentical HSCT, and more prompt requests for biopsy. On the contrary, the tendency to diminish/discontinue immunosuppressive therapy in adult recipients, and to apply preventive/preemptive modulation of immune suppressive therapy based on systematic EBV viral load monitoring in pediatric recipients, may have contributed to the moderation of PTLD incidence [2,3].

26.4 Pathogenesis

Risk factors for PTLD are EBV status at time of transplantation, EBV mismatch, type of transplanted organ (e.g., heart greater than lung greater than liver greater than hematopoietic stem cell greater than pancreas greater than kidney), and type and duration of immunosuppressive regimens (e.g., calcineurin inhibitors [cyclosporine and tacrolimus] greater than antimetabolite mycophenolate mofetil greater than proliferation signaling inhibitors [sirolimus and everolimus] greater than anti-interleukin-2 receptor [CD25] monoclonal antibodies [basiliximab and daclizumab]).

Immune suppressive medication often decreases EBV-specific cytotoxic T-lymphocytes and thus increases uncontrolled proliferation of EBV-infected B cells. In non-EBV-associated PTLD (representing 30% of PTLD case), other viruses (e.g., HHV-8, HCV, and HBV) may play a possible role in late onset PTLD. This leads to the selection of oligoclonal and subsequently monoclonal subpopulations of follicular/postfollicular B-cells or postthymic T-cells, with lowered susceptibility to immune regulation [4].

Molecularly, translocations (e.g., 8q24 [*C-MYC*], 3q27 [*BCL-6*], 14q32 [*IgH, TCL1*]), copy number gains (e.g., 2p24-p25 [*CD138*], 4q21.21, 5p, 7q) and losses (e.g., 1p, 1q [LOH], 2p16.1 [*FRA2E: FANCL, VRK2*], 6q [*PRDM1, TNFAIP3*], 9p [LOH], 17p13 [*TP53*]), gene mutations (e.g., *PIM1, PAX5, C-MYC*, RhoH/TTF), DNA polymorphisms in the host (e.g., *IFN-gamma, IL-10, TGF-beta, TNF-alpha, HLA*) and EBV genome (*LMP1, BZLF1*), and epigenetic alterations (*DAP-k, MGMT, SHP1, TP73*) are implicated in the pathogenesis of B-cell PTLD [5].

26.5 Clinical features

In line with their diverse underlying morphologies and various allografts, PTLD are associated with a range of clinical symptoms—from no symptoms, a mononucleosis-like presentation, to very aggressive disease. The most common symptoms are a painless lump (usually in the neck, armpit, or groin), lymphadenopathy (swollen lymph node), B-symptoms (unexplained fever >38°C, night sweats, unexplained weight loss >10% baseline during the prior 6 months), rapid multi-organ failure, and death (in 25%–60% of cases). Patients with SOT often have extranodal invasion (e.g., gastrointestinal tract 23%–56%, bone marrow 17%, and CNS 5%–13%) with related symptoms.

26.6 Diagnosis

Diagnosis of PTLD relies on observation of clinical symptoms, detection of EBV in peripheral blood, imaging evaluation of nodal and extranodal involvement (using ^{18}F-FDG–PET/CT), and histopathological and immunohistochemical examination of the underlying lymph node/organ.

Depending on cellular constituents, degree of resemblance to reactive or neoplastic lesions known in immunocompetent hosts, and association with the herpesvirus EBV, PTLD is identified as (i) benign polyclonal early lesions (plasmacytic hyperplasia-PTLD, infectious mononucleosis-PTLD, and florid follicular hyperplasia-PTLD); (ii) polymorphic PTLD; (iii) malignant monoclonal monomorphic PTLD; and (iv) classic Hodgkin lymphoma-PTLD (Table 26.1) [1,6].

Diagnosis of early lesions, polymorphic PTLD, and monomorphic PTLD should focus on underlying architecture (partially preserved/destructed/destructed), cell types (plasma cells, small lymphocytes, and immunoblasts/complete spectrum of B cell maturation/fulfilling criteria for lymphoma), immunohistochemistry (no diagnostic value/mixture of B and T cells/mostly CD20 positive), EBV (100%/>90%/70%), clonality (mostly polyclonal/variable/monoclonal), and oncogenic mutations (absent/variable BCL6/oncogenes and tumor suppressor genes).

Differential diagnoses for PTLD include specific and nonspecific lymphoplasmacytic infiltrations associated with infection, graft rejection, graft-versus-host disease, or recurrence from a known lymphoma that developed prior to transplant.

Staging of PTLD requires input from CT scan (abdomen/thorax/pelvis), MRI (brain), or CSF analysis, and bone marrow examination and categorization according to the Ann Arbor classification. Based on the number of involved lymph node regions, the localization of nodal involvement, and the presence of organ invasion, PTLD is classified as stage I to IV, with stage I and II considered as limited disease, with stages III and IV as more advanced or disseminated disease [6].

26.7 Treatment

As the development of PTLD implies a high degree of overimmunosuppression, the mainstay treatment for this disease is to reduce or withdraw immunosuppressive therapy in order to reconstitute the immune system and to restore the EBV-specific T-cell response. This involves discontinuation of antimetabolites and reduction of calcineurin inhibitor dose by 50%. In general, the absence of bulky disease (>7 cm), early stage (Ann Arbor I–II), and lower age (less than 50 years) are important predictors for response to reduction in suppression alone [7,8].

Additional measures include surgical excision or targeted radiation therapy for localized lesions that do not respond to immunosuppression reduction and the use of rituximab (a chimeric monoclonal antibody against CD20) in highly dominant B-cell CD20+ clones that rarely respond to reduction in immunosuppression. However, for aggressive CD20 negative PTLD or PTLD unresponsive to reduction in immunosuppression, anthracycline-based combination chemotherapy (e.g., CHOP [cyclophosphamide, doxorubicin, vincristine, prednisone], PROMACE-CYTABOM [cyclophosphamide, doxorubicin, etoposide, prednisone, cytarabine, bleomycin, vincristine, methotrexate, leucovorin], DHAP [dexamethasone, cytarabine, cisplatin], and VAPEC-B [doxorubicin, etoposide, cyclophosphamide, methotrexate, bleomycin, vincristine]) is necessary [7,8].

Because most anti-EBV compounds target the lytic phase—a minor component of EBV infection in PTLD, they are generally ineffective for the treatment of PTLD.

26.8 Prognosis

PTLD have 3-year and 5-year overall survival rates of 60% and 40%, respectively. Specifically, early lesions have a good prognosis as they tend to regress with reduction in immune suppression. Polymorphic PTLD may also regress, although rejection leading to graft loss and death can occur. Monomorphic PTLD does not usually regress and requires additional therapies (rituximab and/or combination chemotherapy) [6].

Unfavorable prognostic factors for PTLD include an age greater than 60; elevated lactate dehydrogenase; poor performance state; advanced Ann Arbor stage; presence of extranodal localizations; CNS invasion; T-, NK-, or EBV-negative PTLD; and poor response of thoracic organ transplant recipients to rituximab monotherapy.

References

1. Swerdlow SH, Campo E, Pileri SA, et al. The 2016 revision of the World Health Organization classification of lymphoid neoplasms. *Blood.* 2016;127(20):2375.
2. Neuringer IP. Posttransplant lymphoproliferative disease after lung transplantation. *Clin Dev Immunol.* 2013;2013:430209.
3. Dierickx D, Cardinaels N. Posttransplant lymphoproliferative disorders following liver transplantation: Where are we now? *World J Gastroenterol.* 2015;21(39):11034–43.
4. Green M, Michaels MG. Epstein–Barr virus infection and posttransplant lymphoproliferative disorder. *Am J Transplant.* 2013;13 Suppl 3:41–54.
5. Morscio J, Dierickx D, Tousseyn T. Molecular pathogenesis of B-cell posttransplant lymphoproliferative disorder: What do we know so far? *Clin Dev Immunol.* 2013;2013:150835.
6. PathologyOutlines.com website. *Post-transplant lymphoproliferative disorders.* http://www.pathologyoutlines.com/topic/lymphomanonBposttrans.html. Accessed 27 April 2017.
7. Dierickx D, Tousseyn T, Gheysens O. How I treat posttransplant lymphoproliferative disorders. *Blood.* 2015;126(20):2274–83.
8. Kubica MG, Sangle NA. Iatrogenic immunodeficiency-associated lymphoproliferative disorders in transplant and nontransplant settings. *Indian J Pathol Microbiol.* 2016;59(1):6–15.

27
Histiocytic and Dendritic Cell Neoplasms

27.1 Definition

Classified along with mature B-cell neoplasms, mature T and NK neoplasms, Hodgkin lymphoma, and post-transplant lymphoproliferative disorders (PTLD) under the mature lymphoid, histiocytic, and dendritic neoplasms category, histiocytic and dendritic cell neoplasms consist of (i) histiocytic sarcoma (HS), (ii) Langerhans cell histiocytosis (LCH), (iii) Langerhans cell sarcoma (LCS), (iv) indeterminate dendritic cell tumor (IDCT), (v) interdigitating dendritic cell sarcoma (IDCS), (vi) follicular dendritic cell sarcoma (FDCS), (vii) disseminated juvenile xanthogranuloma (DJX), and (viii) Erdheim–Chester disease (ECD) (Table 27.1) [1].

27.2 Biology

Derived from myeloid stem cell/monoblasts (see Chapter 13, Figure 13.1), dendritic cells, monocytes, macrophages, and histiocytes (tissue-resident macrophages) are members of the mononuclear phagocyte system. Macrophages are large ovoid cells mainly involved in the clearance of apoptotic cells, debris, and pathogens; dendritic cells are starry cells that are responsible for presenting antigens on major histocompatibility complex molecules and activating naive T lymphocytes.

Human dendritic cells comprise two groups: plasmacytoid and myeloid. The latter (mDC) are further divided into two subsets: mDC1 (CD141+) and mDC2 (CD1c+). Located within the epidermis, mucosae, or bronchial epithelium, Langerhans cells are dendritic cells that are characterized by their expression of CD1a, possession of Birbeck granules, and their involvement in antigen presentation to T cells upon activation. After activation by local inflammation, Langerhans cells migrate to draining lymph nodes and differentiate into interdigitating cells (IDC) , which present antigens to T cells located in the paracortex and regulate cellular immune. Derived from stromal cells located in the follicles of activated lymph nodes, follicular dendritic cells accumulate and entrap immune complexes and store antigens that serve as a nidus for B-cell proliferation and differentiation with help from T-cells.

Table 27.1 Characteristics of Histiocytic and Dendritic Cell Neoplasms

Type	Biological Features	Affected Organs/ Tissues	Diagnostic Features
Histiocytic sarcoma (HS)	Malignant proliferation of mature tissue histiocytes; expression of histiocytic markers (without dendritic cell markers)	Lymph node (33%), intestines (33%), skin (solitary or multiple lesions 33%)	Effacement of architecture by diffuse, non-cohesive proliferation of large, round/oval, spindling neoplastic cells (pleomorphic nuclei, vesicular chromatin, abundant eosinophilic cytoplasm, variable atypia); immunophenotype: CD163+, CD68+, lysozyme +, CD1a–, CD21–, CD35–
Langerhans cell histiocytosis (LCH)	Neoplastic proliferation of Langerhans cells; evident epidermotropism in intraepidermal microabscesses; unifocal to multisystem disease in ~10%	Bones (skull, femur, pelvis, ribs), lymph node (sinuses, paracortex), skin, lung	Oval Langerhans cells (grooved, folded, indented, or lobulated nuclei, fine chromatin, pale to pinkish cytoplasm, cytologic atypia); eosinophilic infiltration, Birbeck granules; immunophenotype: S100+, CD1a+, CD207 (langerin)+, vimentin+, CD4+, CD30–; IgH gene, TCR α, β, γ chain genes
Langerhans cell sarcoma (LCS)	Neoplastic proliferation of Langerhans cells; *de novo* or progressing from antecedent LCH (more aggressive than LCH)	Lymph nodes, liver, spleen, lung, bone	Nodal architectural effacement by sheets of Langerhans cells (elongated nuclei, prominent nuclear grooves and nucleoli, abundant eosinophilic cytoplasm, high mitoses); sinusoidal pattern in lymph nodes; Birbeck granules; immunophenotype: S100+, CD1a+, CD207 (langerin)+, CD4+
Indeterminate dendritic cell tumor (IDCT)	Expansion of indeterminate cells (possibly mature Langerhans cells without Birbeck granules/langerin expression	Skin	Solid sheets of indeterminate cells (indistinct cell borders, irregular nuclear grooves and clefts, inconspicuous nucleoli, abundant eosinophilic cytoplasm) in the dermis and subcutis but not epidermis; immunophenotype: S100+, CD1a+, CD207 (langerin)–, CD21–, CD23–, CD35–, B/T cell markers negative

(Continued)

Table 27.1 ***(Continued)*** **Characteristics of Histiocytic and Dendritic Cell Neoplasms**

Type	Biological Features	Affected Organs/ Tissues	Diagnostic Features
Interdigitating dendritic cell sarcoma (IDCS)	Neoplastic proliferation of spindle to ovoid cells with phenotypic features of IDC	Solitary lymph node, skin, intestine, soft tissue	Paracortical distribution with residual follicles; whorls of large spindle to ovoid cells (indistinct cell borders, vesicular chromatin, abundant eosinophilic cytoplasm); immunophenotype: CD21–, CD23–, CD1a–, vimentin–, S100+, CD45+, CD68+/–
Follicular dendritic cell sarcoma (FDCS)	Neoplastic proliferation of FDC; association with Castleman disease in 10%–20% of cases	Lymph node, extranodal (tonsil, spleen, oral cavity, GI tract, liver, soft tissue, skin, breast)	Fascicles, whorls, diffuse sheets or nodules of spindled to ovoid cells (bi- and multinucleated, nuclear pseudo-inclusions, cytological atypia, mitotic figures, coagulative necrosis); immunophenotype: CD21+, CD23+, CD35+, C68+/–, CD1a–
Disseminated juvenile xanthogranuloma (DXJ)	Cutaneous lesion is benign and often regresses; CNS involvement causes morbidity and mortality	Skin, mucosa, CNS	Small and oval cells (bland round/oval nucleus without grooves, pink cytoplasm); immunophenotype: vimentin+, lysozyme+, CD14+, CD68+, stabilin1+, CD163+, factor XIIIa+, CD1a–
Erdheim–Chester disease (ECD)	Expansion of EDC cells; 20% patients with LCH lesions; extracutaneous or disseminated JXG with gain-of-function mutations	Skeletal (>95%), cardiovascular (50%), kidneys/ureters (33%), CNS (20%), eyelid (xanthelasma)	Infiltration of foamy mononucleated histiocytes with small nucleus; abundant fibrosis, lymphocytes, plasma cells, and neutrophils; immunophenotype: CD68+, CD163+, CD1a–; gain-of-function mutation of BRAF, NRAS, KRAS, or MAP2K1

LCH and LCS result from clonal proliferation of Langerhans cells; HS, DJX, and ECD are due to an accumulation of histiocytes (tissue-resident macrophages); INDCT, IDCS, and FDCS arise from clonal expansion of antigen-presenting dendritic cells.

Collectively, LCH, LCS, IDCS, FDCS, and IDCT are referred to as dendritic cell sarcomas (DCS), of which FDCS is the most common subtype followed by IDCS. Although most DCS originate from cervical, mediastinal, axillary, and inguinal lymph nodes, some also have extranodal involvement. Similar to other low-grade soft tissue sarcomas, DCS demonstrates a 30% overall risk for developing local recurrences and metastases. It is believed that HS, LCH, and IDCS evolve from bone marrow precursors; whereas FDCS, IDCT, and DJX originate from stromal/mesenchymal precursors [2,3].

27.3 Epidemiology

Dendritic and histiocytic cell neoplasms are rare malignancies, making up <1% of all neoplasms of lymph nodes or extramedullary locations.

HS has a median age of 46 years and a slight male preference. LCH shows an incidence of five cases per million and affects mostly infants and young children with a male to female ratio of 3.7:1. LCS has a median age of 41 years and a predominantly female distribution. IDCT tends to occur in adults without preference for either sex. IDCS mainly affects elderly adults without sex predilection. FDCS demonstrates a median age of 44 years and a benign course, with a median survival of 168 months. DXG preferentially affects children, with 50% of cases occurring in the first year of life and 20% in adolescents and young adults, and a slight male predilection (male to female ratio of 1.1–1.4:1). As an adult form of DXG, ECD shows a mean age at diagnosis of 55 to 60 years, and a male to female ratio of 3:1.

27.4 Pathogenesis

The etiology of dendritic and histiocytic cell neoplasms is currently unknown, although some types harbor BRAF(V600E) mutations (HS 62.5%, ECD 54%, LCH 38%, and FDCS 18.5%).

The *braf* gene is a proto-oncogene encoding a 766-amino acid serine/threonine protein kinase (BRAF) belonging to the Raf kinase family. BRAF phosphorylates and activates MEK1/2 and plays a key role in the regulation of the MEKK/FRK pathway, which affects cell differentiation, proliferation, growth, and apoptosis. Mutations in the *braf* gene can be inherited and cause birth defects (cardiofaciocutaneous syndrome—a disease characterized by heart defects, mental retardation, and a distinctive facial appearance); it can also appear later in life and act as an oncogene, leading to non-Hodgkin lymphoma, colorectal cancer, malignant melanoma, papillary thyroid carcinoma, non-small-cell lung carcinoma, pulmonary adenocarcinoma, brain tumors (e.g., glioblastoma and pilocytic astrocytoma), and inflammatory diseases (e.g., ECD). Of more than 40 different mutations identified in the *braf* gene, the most predominant one is a missense mutation involving a thymidine to adenosine transversion at nucleotide 1,799 located in exon 15, which results in the substitution of valine with glutamic acid at amino acid 600—that is, BRAF(V600E) [4].

27.5 Clinical features

Histiocytic sarcoma (HS) is a rare non-Langerhans histiocyte disorder of mature tissue histiocytes that is associated with systemic symptoms (fever and weight loss), skin manifestations (benign-appearing rash, solitary lesions, or innumerable tumors on trunk and extremities), GI lesions (possible intestinal obstruction), hepatosplenomegaly, lytic lesions (bone), and pancytopenia (bone marrow).

Langerhans cell histiocytosis (LCH) often appears as a unifocal disease (lytic bone lesions or diaphysis with erosion into soft tissues) in older children and adults; a multifocal, unisystem disease (multiple destructive bone lesions, adjacent soft tissue masses; skull often involved with exophthalmos, diabetes insipidus, and tooth loss) in young children; and a multifocal, multisystem disease (fever, skin manifestations, hepatosplenomegaly, lymphadenopathy, bone lesions, pancytopenia in infants). LCH in adult lungs often presents with innumerable bilateral nodules <2.0 cm in diameter.

Langerhans cell sarcoma (LCS) is often an extranodal tumor with skin and bone involvement and with multi-organ involvement of the lymph nodes, lungs, liver, and spleen. Clinical presentations may consist of pancytopenia, generalized lymphadenopathy, and mild splenomegaly.

Indeterminate dendritic cell tumor (IDCT) is restricted to the skin without systemic symptoms. IDCS usually forms an asymptomatic mass, although systemic disease may show fatigue, fever, night sweats, and hepatosplenomegaly.

Follicular dendritic cell sarcoma (FDCS) usually manifests as a slow-growing mass in the head and neck or abdominal lymph nodes.

Erdheim-Chester disease (ECD) represents an extracutaneous or disseminated JXG with gain-of-function mutations. It often has a heterogeneous clinical course, with some patients showing few symptoms and others having progressive and lethal disease. It may cause bilateral, symmetric cortical osteosclerosis of the diaphyseal and metaphyseal regions (more than 95% of cases), along with cardiovascular signs (50%), retroperitoneal fibrosis (the kidneys and ureters), and cutaneous manifestation. In 20% of patients, LCH lesions may appear.

27.6 Diagnosis

Tumors of histiocytic and dendritic cell origin are diagnosed through histology and immunohistochemistry (Table 27.1).

Differential diagnoses for HS include diffuse large B-cell lymphoma (diffuse histiocytic lymphoma), anaplastic large cell lymphoma (histiocytic medullary reticulosis and malignant histiocytosis), hemophagocytic syndrome, hemophagocytic lymphohistiocytosis, LCH, metastatic carcinoma, or melanoma.

Differential diagnoses for IDCT are LCH, pityriasis rosea, scabies, and T-cell lymphomas (cutaneous T-cell hyperplasia and mycosis fungoides).

Differential diagnoses for FDCS and IDCS include anaplastic large cell lymphoma, inflammatory pseudotumors, intranodal myofibroblastoma, LCH, non-Hodgkin lymphoma, peripheral nerve sheath tumors, and true histiocytic lymphomas.

Differential diagnoses for JXG are LCH, fibrohistocytic lesion NOS, reticulohistiocytoma, hemangioendothelioma, Spitz nevus, malignant fibrous histiocytoma, rhabdomyosarcoma, dermatofibroma, eruptive xanthomas, mastocytoma, papular xanthoma, tuberous xanthoma, and xanthoma disseminatum.

27.7 Treatment

Treatment options for histiocytic and dendritic cell neoplasms consist of surgical resection, chemotherapy, and/or radiotherapy [5,6].

For HS, surgical resection is the mainstay of treatment, while adjuvant radiotherapy may help reduce local recurrence rates. For disseminated HS, lymphoma-based treatments (e.g., cyclophosphamide, doxorubicin, vincristine, prednisone [CHOP] or CHOP and etoposide) should be considered.

Surgical resection, skin-directed therapy, or radiotherapy may be used for a majority of LCH cases. For systemic LCH, agents (e.g., methotrexate, cyclophosphamide, cyclosporine, 6-mercaptopurine, and vinblastine) may be administered, and progressive disease or recurrence can be treated with

2-chlorodeoxyadenosine and cytarabine. An inhibitor of mutated BRAF (vemurafenib) may be applied for patients with LCH carrying $BRAF^{V600E}$ mutations.

Localized treatment of IDCT leads to complete remission. Surgical resection is recommended for localized IDCS, although many IDCS lesions can spontaneously regress.

Combination chemotherapy (CHOP; ifosfamide, carboplatin, and etoposide [ICE]; and adriamycin, bleomycin, vinblastine, and dexamethasone [ABVD]) and radiotherapy are useful for treating advanced FDCS.

No treatment is needed for localized asymptomatic JXG.

27.8 Prognosis

As a neoplastic proliferation with morphological and immunophenotypic features of mature tissue histiocytes, HS has a poor prognosis due to rapid progression and low response to therapy, with a median survival of 5 months for HS in the CNS. Multisystem disease or tumors >3.5 cm carry the worst prognosis for HS. LCH has a greater than 95% survival in unifocal disease and 75% survival when two organs are involved; LCH of the lungs can regress after the cessation of smoking. LCS is an aggressive neoplasm with an overall survival rate of 50%. FDCS has 2-year survival rates of 84.2%, 80%, and 42.8% for early, locally advanced, and distant metastatic disease, respectively. JXG with only skin or soft tissue involvement has a high survival rate, with lesions spontaneously disappearing over time in a majority of cases.

References

1. Swerdlow SH, Campo E, Pileri SA, et al. The 2016 revision of the World Health Organization classification of lymphoid neoplasms. *Blood.* 2016;127(20):2375.
2. Dalia S, Shao H, Sagatys E, Cualing H, Sokol L. Dendritic cell and histiocytic neoplasms: Biology, diagnosis, and treatment. *Canc Control.* 2014;21(4):290–300.
3. Emile JF, Abla O, Fraitag S, et al. Revised classification of histiocytoses and neoplasms of the macrophage-dendritic cell lineages. *Blood.* 2016;127(22):2672–81.
4. Arceci RJ. Biological and therapeutic implications of the BRAF pathway in histiocytic disorders. *Am Soc Clin Oncol Educ Book.* 2014:e441–5.

5. Gounder M, Desai V, Kuk D, et al. Impact of surgery, radiation and systemic therapy on the outcomes of patients with dendritic cell and histiocytic sarcomas. *Eur J Canc*. 2015;51(16):2413–22.
6. Dalia S, Jaglal M, Chervenick P, Cualing H, Sokol L. Clinicopathologic characteristics and outcomes of histiocytic and dendritic cell neoplasms: The Moffitt cancer center experience over the last twenty five years. *Cancers* (*Basel*). 2014;6(4):2275–95.

28 Thymoma and Thymic Carcinoma

Francesco Facciolo, Mirella Marino, and Maria Teresa Ramieri

28.1 Definition

Thymoma and thymic carcinomas (TC) are tumors arising in the antero-superior mediastinum from the epithelial compartment of the thymus gland and are collectively also called "thymic epithelial tumors" (TET). These tumors have different biological and clinical characteristics and background cells, with the thymomas (THY) being composed of epithelial cells of spindle or polygonal/dendritic shapes as well as a variable percentage of (predominantly) T lymphocytes. TC, indeed, are morphologically similar to carcinomas arising in other sites and organs, and several variants are recognized, with squamous cell carcinoma (SCC) being the most frequent subtype. Also, neuroendocrine tumors of thymic derivation (NETT) with variable differentiation degrees and malignancy characteristics occur. Table 28.1 presents the 2015 World Health Organization (WHO) classification of TET [1] along with the associated International Classification of Diseases for Oncology (ICD-O) morphological codes.

THY are the most frequent among TET, and their classification has been debated among pathologists for several years. The WHO classification utilizes letters A and B to indicate THY subtypes and their epithelial cell arrangement (Table 28.1). The present WHO classification, based on the morphology of epithelial cells (spindle and/or polygonal/dendritic), aggregation pattern, and degree of lymphocytic infiltration, is now widely accepted, and recently refined criteria have contributed to the clarification of controversial diagnostic points [2]. The 2015 WHO classification and its previous 1999 and 2004 editions, in the author's opinion, reflect in some way the resemblance of THY to the normal thymic architecture that forms the basis of the "histogenetic thymoma classification" [3].

The lymphocytes in THY are not neoplastic, demonstrating an immature cortical thymocyte phenotype as well as the capacity to undergo development and maturation (even if defective) in the tumor microenvironment before being exported to the periphery. The highly specialized microenvironment built up in THY and the strict correlations among thymoma epithelial cells and the intratumorous lymphocytes set the basis for the immune dysregulation

Table 28.1 WHO Classification 2015 of Thymic Epithelial Tumors (TET)

Epithelial Tumors	NUT Carcinoma
Thymoma	Undifferentiated carcinoma
Type A thymoma (including atypical variant)	Other rare thymic carcinomas
Type AB thymoma	Adenosquamous carcinoma
Type B1 thymoma	Hepatoid carcinoma
Type B2 thymoma	Thymic carcinoma, NOS
Type B3 thymoma	Thymic neuroendocrine tumors
Micronodular thymoma with lymphoid stroma	Carcinoid tumors
Metaplastic thymoma	Typical carcinoid
Other rare thymomas	Atypical carcinoid
Microscopic thymoma	Large cell neuroendocrine carcinoma
Sclerosing thymoma	Combined large cell neuroendocrine carcinoma
Lipofibroadenoma	Small cell carcinoma
Thymic carcinoma	Combined small cell neuroendocrine carcinoma
Squamous cell carcinoma	Combined small cell carcinoma
Basaloid carcinoma	
Mucoepidermoid carcinoma	
Lymphoepithelioma-like carcinoma	
Clear cell carcinoma	
Sarcomatoid carcinoma	
Adenocarcinomas	
Papillary adenocarcinoma	
Thymic carcinoma with adenoid cystic	
carcinoma-like features	
Mucinous adenocarcinoma	
Adenocarcinoma NOS	

Sources: Travis WD., et al. *WHO Classification of Tumours of the Lung* (4th ed), Pleura, Thymus and Heart. Lyon: IARC press, 2015 and Fritz A., et al. *International Classification of Disease for Oncology* (3rd ed), World Health Organization, Geneva, 2000.

Note: According to the ICD-O-3 classification the following morphologic codes were associated with TET: malignant thymoma (8580–8586; thus including not otherwise specified [NOS, 8580], type A [8581], type AB [8582]); type B (8583, 8584, 8585); type C (8586); squamous cell carcinoma (8051–8076, 8078, 8083–8084); undifferentiated carcinoma (8020–8022); lympho-epithelial carcinoma (8082); and adenocarcinoma.

and the autoimmune diseases that often are THY-associated. Thymoma and thymic carcinomas lack any resemblance to the normal thymic architecture and are largely invasive, with frequent invasion into the mediastinum, the lung, the pleura, and intrathoracic great vessels. Thymic neuroendocrine tumors include low-, intermediate-, and high-grade variants morphologically resembling their lung counterparts. Paraneoplastic endocrine syndromes are often linked to well-differentiated forms.

28.2 Biology

THY are slowly growing tumors. Previously considered almost always "benign" when removed in early stages, THY frequently give local relapses and intrathoracic disseminations (most often in the pleura). The occurrence of metastasis or recurrence is observed in 31% of TET cases, with half of these cases also showing three or more recurrences or metastases during their oncological history [4]. Occurrence in the lungs, rare intra thoracic lymph node metastases, and extrathoracic spread are related to stage and histologic subtypes. The development of extrathoracic metastases is a rare event in THY, mainly occurring in cervical lymph nodes, bone marrow, or (rarely) in the liver. Thymoma and thymic carcinomas (TC) are highly aggressive neoplasms; in addition to a local invasive tendency, they show an extrathoracic metastatic tendency toward the liver, bone marrow, and other sites. Therefore, THY should be considered neoplasms with an uncertain biological propensity to recur and to give intrathoracic metastases, whereas TC are by definition aggressive tumors. NETT are frequently locally invasive and may also metastasize in extrathoracic sites, with variable frequency depending on their subtype.

The tendency of THY to recur after R0 (no residual tumor tissue) resection is probably related to thymic nests/remnants in the mediastinum; hence, the adequate surgical management is based on total thymectomy even for relatively limited tumors. Moreover, the possibility of new clones emerging cannot be excluded. A molecular classification of TET would contribute to identify prognostic as well as predictive biomarkers and to promote personalized treatment approaches. Use of next-generation sequencing (NGS) led to the identification in a series of 274 TET of a GTF2I gene mutation in 82% of type A and 74% of type AB thymomas (two relatively indolent thymoma subtypes) but rarely in the aggressive subtypes [5]. The GTF2I gene, situated on chromosome 7, encodes a phosphoprotein functioning as a regulator of transcription. The missense mutation in GTF2I (in chromosome 7 c.74146970T>A) correlated with better survival. Moreover, a higher number of mutations in TC than in THY and recurrent mutations of known cancer genes, such as TP53, CYLD, CDKN2A, BAP1, and PBRM1 were identified [5].

28.3 Epidemiology

TET, including THY and TC, are rare tumors, accounting for 0.13 in 100,000 in the United States according to the National Cancer Institute's Surveillance, Epidemiology, and End Results (SEER) program/SEER database. THY are uncommon in children and young adults, rise in incidence in middle age, and peak in the seventh decade [6]. In Europe, population-based data from different European cancer registries (CRs) participating in the RARECARE project gave an incidence rate of 0.17 in 100,000, with "malignant" thymoma accounting for 0.14 in 100,000. THY are far more frequent than TC. The latter have an incidence rate of 0.2 to 0.5 per million individuals. TET has lowest incidence in Northern and Eastern Europe, UK and Ireland, but somewhat higher incidence in Central and Southern Europe. In the United States, the incidence of THY is higher in blacks and especially in Asians/Pacific Islanders than among whites or Hispanics [6]. There are no or only limited differences in the incidence between the sexes, with females prevailing in THY subtypes A, AB, and B1 and males in TC. The site of TET occurrence is the antero-superior mediastinum at the site of thymic development. However, ectopic tumors have been described in the thorax mainly in the pleura and in the lung or in other mediastinal areas—particularly in the aortopulmonary window and retrocardiac area. Outside the thorax, ectopic TET predominantly occur along the route followed by the thymic primordium during its descent to the mediastinum from the third branchial cleft (i.e., in the neck and in the thoracic inlet). THY are associated with an increased risk of second cancer development, including both frequent tumors as well as rare malignancies, with the most frequently associated tumors being non-Hodgkin lymphoma [6].

28.4 Pathogenesis

No environmental or viral risk factor is currently known to play a role in TET development. A genetic basis is recognized for very rare cases occurring in the setting of multiple endocrine neoplasia 1 (MEN1) syndrome. A nine gene-based signature, including three upregulated genes (AKR1B10, JPH1, and NGB), was found to predict the metastatic behavior. Recently, a large microRNA cluster on chromosome 19q13.42, constituting a transcriptional hallmark, was noted in all A and AB tumors. This microRNA cluster, virtually absent in the other THY and normal tissues and overexpressed in A and AB tumors, activates the PI3K/AKT/mTOR pathway. This suggests the potential use of PI3K inhibitors in patients with these tumor subtypes. Moreover, a preliminary mature microRNA signature, detectable in formalin-fixed paraffin embedded (FFPE) tissues, has been identified by microarray and bioinformatics analysis. Distinct groups of miRNAs appear to express differentially expressed among TET and normal thymic tissues and among THY and TC and histotype groups. Correlated putative molecular pathways targeted by these differentially miRNAs include pathways related to cell adhesion/motility and additional pathways related to cancer phenotypes such as Wnt, Notch, TGF-α, ErbB, p53, VEGF, and MAPK signaling pathways. To achieve a comprehensive understanding of the TET genome, the National Cancer Institute (NCI) in the United States promotes the Cancer Genome Atlas-THY (TCGA-THY) in the framework of rare tumor genomic studies, and the results are expected soon.

28.5 Clinical features

About 50% of thymic tumors are asymptomatic until achieving a large size. Others produce local (pain, cough, respiratory difficulties, and tachycardia) or systemic (fever and weight loss) symptoms. Symptomatic cases are most likely to have a malignant character. In some cases, myasthenic symptoms drive the attention to a mediastinal mass. Very rarely, thymic tumor is revealed by an extrathoracic metastasis, usually causing diagnostic difficulties if the primary tumor is unknown or underevaluated. In addition to myasthenia gravis (MG), which is the main TET-associated autoimmune disease (occurring in 15%–20% of THY, but not in TC), other possible autoimmune disorder associations include neuromuscular, rheumatic, and vasculitic disorders as well as hematopoietic (e.g., pure red cell aplasia, PRCA) and dermatologic diseases. Bacterial and opportunistic infections may occur, related to hypogammaglobulinemia (Good's syndrome, a severe acquired immunodeficiency disease). Among the myasthenic symptoms, asymmetric palpebral ptosis, diplopia, "bulbar" symptoms—eventually associated to weakness in some muscle groups, and respiratory weakness are early MG symptoms. Autoimmune symptoms could manifest as early signs of TET, after tumor removal, or even several years after the removal, revealing a recurrence.

28.6 Diagnosis

CT scan is the gold standard in the diagnostic workup of mediastinal masses, whereas 18-fluorodeoxyglucose PET is only appropriate for staging purposes in advanced tumors. The workup of patients with mediastinal masses should consider the demographical characteristics, symptoms, imaging features, and blood and serological findings of the patients. Electromyography (EMG) to reveal a latent MG should be part of a preoperative workup of patients with a mediastinal mass of suspicious thymic origin. In THY patients, MG is nearly invariably associated with the presence of serum anti-acetylcholine receptor antibodies (AChR-Abs); in a high proportion of THY-MG cases, Abs against the muscle giant protein titin, previously identified as "striational" muscle Abs, and anti-ryanodine receptors are also found. Occult MG might cause a myasthenic crisis following tumor removal, therefore pointing to the need for preoperative diagnostic assessment of a clinically latent MG. The diagnosis of TET can be suspected on the basis of the CT imaging features and, in most cases, a radical surgical approach can be planned without a diagnostic biopsy. However when unresectability is suspected, a core biopsy or a surgical biopsy obtained by board certified thoracic surgeon is required to define the diagnosis by pathologist experienced in rare thoracic tumor/thymic pathology (Figure 28.1). Very severe clinical consequences could derive if the tumor is treated on the basis of an incorrect diagnosis.

Differential diagnosis: TET are the most frequent primary mediastinal tumors in adults (35% of cases); however, they have to be distinguished from lymphomas occurring in the mediastinum (of both Hodgkin and non-Hodgkin types), which account for 25% of cases and (specifically in males) from germ cell tumors (GCTs; 20%). THY may also arise primarily in the lung; on the contrary, lung carcinoma metastatic (in the mediastinum) may simulate a primary TC (mainly of the squamous type). The possible occurrence of TET in ectopic sites has been already mentioned in Section 28.3.

28.7 Treatment

Due to the anatomical location and the clinical complexity of TET, a multidisciplinary tumor board discussion pre- and postoperatively is advisable. Tumor resectability drives the treatment strategy, with surgery being the gold standard for most cases—alone or in conjunction with other treatment options. Open thymectomy (OT) through median sternotomy (with or without an accompanying cervical incision) represents the most widely used approach. Minimally invasive techniques (MITs), including transcervical, video-assisted thoracoscopic (VATS), and robotic video-assisted thoracoscopic (R-VATS) approaches, should be considered in clinical stages I–II, whereas in suspected stage III, not enough data on long-term follow-up after MIT are available. In a metanalysis based review, 62% of thymomas resected by MIT were stage I, while 36% and 2% were stages II and III, respectively. Likewise, 58% of thymomas resected by OT were stage I, and 38% and 5% were stages II and III, respectively. Larger tumors need to be treated by OT. Radical thymectomy with removal of involved structures is indicated for locally invasive tumors whenever feasible. Recurrent TET are also treated by surgery. The role of postoperative radiotherapy is controversial in stage II TC and in "high risk" THY (or B3 histotype or extensive capsular infiltration); in stages III and IVA, radiotherapy is also indicated, although this has been debated. Induction chemotherapy should be given in clinical stage III/IVA and for inoperable and relapsing TET [7,8]. Several excellent reviews are available elsewhere concerning the treatment strategies, risk and stage adapted therapy principles, morbidity and outcome related to the surgical approaches, and targeted therapies for thymic epithelial malignancies.

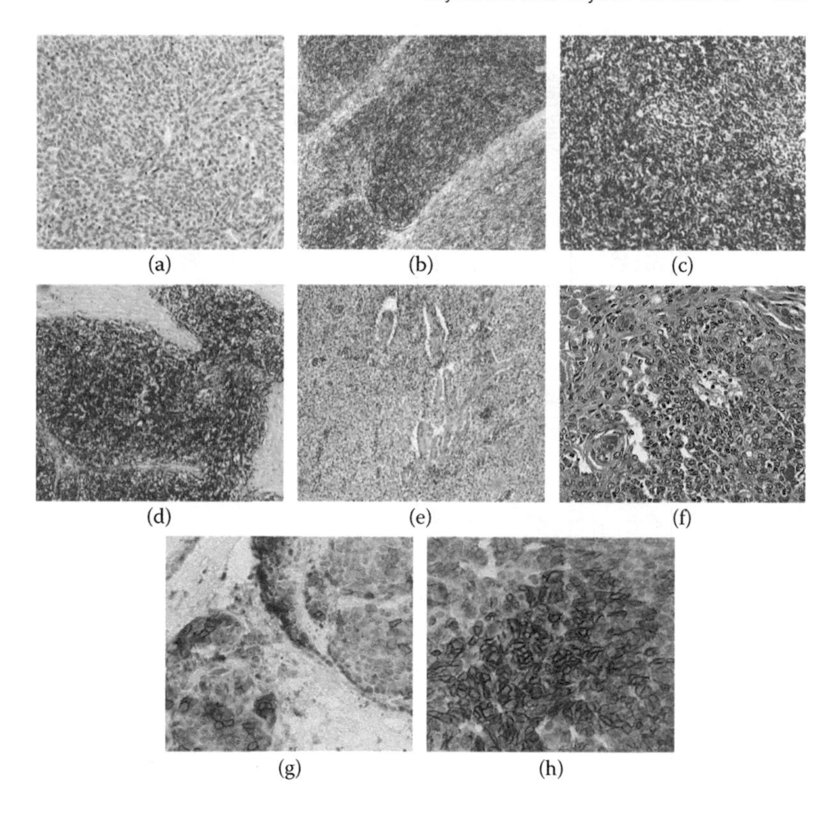

Figure 28.1 (a) *A type thymoma*: classical spindle epithelial cell pattern and paucity of interspersed lymphocytes; (b) *AB type thymoma* showing spindle epithelial cell rich areas, devoid of lymphocytes (bottom right) alternated with lymphocyte-rich area (in the center); histiocytes with clear cytoplasm confer a "starry sky" aspect to the "cortical" area; (c) *B1 type thymoma* have thymus-like architecture, with a predominance of cortical areas (on the left) characterized by a high lymphocytic content conferring the dark blue aspect; pale areas constitute foci of medullary differentiation (so-called medullary islands, to the right) are always present; in the dark blue areas scattered epithelial cell and histiocytes confer a" mottled" appearance; (d) *B2 type thymoma* at low magnification appears lobulated and blue due to the abundance of lymphocytes; at variance from lymphomas a fibrous tumour capsule and lobular architecture are seen; (e) *B3 type thymoma* at low magnification appear usually poorly circumscribed, with typical tumoural lobules separated by fibrous septa; perivascular spaces contain T lymphocytes, a typical morphological feature of B type thymoma; (f) *Thymic carcinoma;* frankly atypical cells with obvious squamous differentiation: a greater amount of eosinophilic cytoplasm and the presence of intercellular bridges; (g) *TC, CD5*: CD5 and CD117 are frequently expressed in thymic carcinomas, and therefore may be of value in the differential diagnosis of thymic TC versus carcinomas of other origin; (h) *TC, CD117*: CD117 is expressed in 78% of thymic carcinoma and in 84% of squamous cell thymic carcinoma.

Table 28.2 Demographical and Clinicopathological Findings and Outcome Indicators of the Main

THY Subtype	Frequency (%)[a]	Masaoka-Koga Stage[b] (%)					OS (WHO) %			Recurrence Rate(%)[c]		Metastasis
		I	II	III	IVA	IVB	5 years	10 years	20 years	5 years	10 years	
Type A	11	60	31	8	<1	<1	100	100		5	9	Exceptional
Type AB	28	67	26	6	1	–	80–100			3		Very rare
Type B1	17	50	37	9	3	1		85–100		11	14	Rare
Type B2	26	32	29	28	8	3		70–90	59–78	14	32	Frequent
Type B3	16	19	36	27	15	3		50–70	25–36	23	29	Frequent

Source: Modified from Ried, M., et al., *Eur. J. Cardiothorac. Surg.*, 49(6), 1545–1552, 2016; data from Travis, W.D., et al., *WHO Classification of Tumours of the Lung, Pleura, Thymus and Heart* (4th ed.), IARC Press, Lyon, 2015; Detterbeck, F.C., et al., *J. Thorac. Oncol.*, 9(9 Suppl 2), S65–S72, 2014; Weis, C.A., et al., *J. Thorac. Oncol.*, 10(2), 367–372, 2015; and authors' own findings.

[a] According to WHO classification in Ref. [1].

[b] As described by Detterbeck et al. [7].

[c] After R0 resection according to ITMIG DB in Ref. [10].

28.8 Prognosis

As radical surgical removal is the gold standard for TET treatment, based on relapse-free survival (RFS) and overall survival (OS) data, the achievement of an R0 resection status by radical surgery is the main prognostic factor. For routine management, TET staging is presently performed according to the Masaoka-Koga system. This staging system is mainly based on post-surgery findings from tumor specimens. Staging according to Masaoka Koga has a significant impact on OS among patients with R0 THY resection. Recently, the International Thymic Malignancy Interest Group (ITMIG), using data derived from a world-wide-based surgical database (DB), has proposed The Union for International Cancer Control (UICC) as a new staging system of TET [9], different from the Masaoka-Koga system. International Thymic Malignancy Interest Group (ITMIG) has proposed that anterior mediastinal nodes be routinely removed along with the thymus and has encouraged a systematic sampling of deep nodes when resecting invasive THY. In a recent report based on the ITMIG DB data, the recurrence rate of THY was found to be strongly correlated with the histotype; moreover, there was a strong association limited to stage I+II between stage and histotype only significant for type AB versus B1, B2, or B3 thymoma [10]. Table 28.2 summarizes the main correlations among TET subtype, Masaoka-Koga staging, and outcome indicators according to WHO data and to recent reports.

References

1. Travis WD, Brambilla E, Burke AP, Marx A, Nicholson AG. *WHO classification of tumours of the lung, pleura, thymus and heart.* (4th ed.). Lyon: IARC Press; 2015.
2. Marx A, Ströbel P, Badve SS, et al. ITMIG consensus statement on the use of the WHO histological classification of thymoma and thymic carcinoma: Refined definitions, histological criteria, and reporting. *J Thorac Oncol.* 2014;9(5):596–611.
3. Marino M, Müller-Hermelink HK. Thymoma and thymic carcinoma. Relation of thymoma epithelial cells to the cortical and medullary differentiation of thymus. *Virchows Arch A Pathol Anat Histopathol.* 1985;407(2):119–49.
4. Khandelwal A, Sholl LM, Araki T, et al. Patterns of metastasis and recurrence in thymic epithelial tumours: Longitudinal imaging review in correlation with histological subtypes. *Clin Radiol.* 2016;71(10):1010–7.
5. Petrini I, Meltzer PS, Kim IK, et al. A specific missense mutation in GTF2I occurs at high frequency in thymic epithelial tumors. *Nat Genet.* 2014;46(8):844–9.

6. Engels EA. Epidemiology of thymoma and associated malignancies. *J Thorac Oncol.* 2010;5(10 Suppl 4):S260–5.
7. Weis CA, Yao X, Deng Y, et al. The impact of thymoma histotype on prognosis in a worldwide database. *J Thorac Oncol.* 2015;10(2):367–72.
8. Ried M, Marx A, Götz A, et al. State of the art: Diagnostic tools and innovative therapies for treatment of advanced thymoma and thymic carcinoma. *Eur J Cardiothorac Surg.* 2016;49(6):1545–52.
9. Detterbeck FC, Stratton K, Giroux D, et al. The IASLC/ITMIG thymic epithelial tumors staging project: Proposal for an evidence-based stage classification system for the forthcoming (8th) edition of the TNM classification of malignant tumors. *J Thorac Oncol.* 2014;9(9 Suppl 2):S65–72.
10. Scorsetti M, Leo F, Trama A, et al. Thymoma and thymic carcinomas. *Crit Rev Oncol Hematol.* 2016;99:332–50.

Glossary

Anaplasia: A term used to describe cancer cells with a total lack of differentiation and with resemblance to original cells either in functions or structures or both; also known as dedifferentiation (backward differentiation).

Aneuploidy: The presence of an abnormal number of chromosomes in a cell (see Diploidy).

Angiogenesis: The growth of new capillary blood vessels from preexisting vessels in the body (especially around a developing neoplasm).

Apoptosis: Programmed cell death, with the cells damaged beyond repair and typically dying swollen and burst, spilling their contents over their neighbors.

Atypia: The state of being not typical or normal. In medicine, atypia is an abnormality in cells, which may or may not be a precancerous indication associated with later malignancy.

Autosomal dominant inheritance: When a dominant gene located on one of the 22 nonsex chromosomes (autosomes) from one parent is mutated, there is a 50% chance that an affected child will inherit the mutated gene (dominant gene) from this parent and a normal gene (recessive gene) from other parent, whereas an unaffected child will have two normal genes (recessive genes) and will not develop or pass on the condition.

Autosomal recessive inheritance: Unlike autosomal dominant inheritance, which involves passing of a mutated gene from one parent, autosomal recessive inheritance involves passing of two mutated genes, one from each parent. Therefore, two carriers each with one mutated gene (recessive gene) and one normal gene (dominant gene) have a 25% chance of getting an unaffected child with two normal genes, a 50% chance of getting an unaffected carrier child, and a 25% chance of getting an affected child with two recessive genes.

Benign tumor: A slow-growing, non-cancerous tumor that does not invade nearby tissue or spread to other parts of the body. In most cases, a benign tumor has a favorable outcome, with or without surgical removal. However, a benign tumor in vital structures such as nerves and blood vessels, or undergoing malignant transformation, often has serious consequence (see Hamartoma).

Biopsy: A procedure to remove tumor tissue or cells or tissues for microscopic examination. This is usually conducted through excisional biopsy (removal of an entire lump of tissue), incisional biopsy (removal of part of a lump or a sample of tissue), core biopsy (removal of tissue using a wide needle), or fine needle aspiration (FNA) biopsy (removal of tissue or fluid using a thin needle).

Blasts, blast cells: Immature cells that give rise to specialized cells. For example, neuroblasts give rise to nerve cells; adipoblasts give rise to adipocytes (fat cells); lymphoblasts give rise to B- and T-lymphocytes; and erythroblasts, myeloblasts, monoblasts, and megakaryoblasts give rise to red blood cells (erythrocytes), white blood cells (eosinophils, basophils, neutrophils, mast cells, and monocytes, but not lymphocytes), and platelets. Out of all blast cells, myeloblasts are frequently referred to as "blasts" (see Hematopoiesis and Lymphopoiesis).

Bone marrow: Forming the soft inner part of some bones (e.g., the skull, shoulder blades, ribs, pelvis, and backbones), bone marrow consists of blood stem cells, mature blood-forming cells, fat cells, and supporting tissues and is largely responsible for producing blood cells after birth while the spleen does so during fetal life (see Hematopoiesis and Lymphopoiesis).

Calcification: The accumulation of calcium salts (e.g., and calcium phosphate) in body tissues such as tumor mass, where they do not usually appear. This leads to tissue hardening and produces a dense opacity on a radiographic image.

Cancer: Plural cancers or cancer, a group of diseases involving uncontrolled expansion of abnormal cells that have the potential to invade nearby tissue and/or spread to other parts of the body via the bloodstream or lymphatic vessels (see Tumor, Neoplasm, and Lesion).

Carcinoma: A type of cancer that begins in a tissue (called epithelium) that lines the inner or outer surfaces of the body.

CT: Computerized tomography (also known as a computed tomography scan [CT scan] or computerized axial tomography [CAT]) utilizes an X-ray machine linked to a computer together with a dye (swallowed or injected into a vein) to take a series of detailed pictures of affected organs or tissues in the body from different angles in order to determine the precise location and dimension of a tumor.

Cyst: A closed capsule or sac-like structure usually filled with a liquid, semi-solid, or gaseous material (but not pus, which is considered an abscess). As an abnormal formation, a cyst on the skin, mucous membranes, and inside palpable organs can be felt as a lump or bump, which may be painless or painful. Although cysts due to infectious causes are preventable, those due to genetic and other causes are not. Most cysts are benign (noncancerous).

Dedifferentiation: See Anaplasia.

Desmoplasia: The growth of fibrous or connective tissue around a neoplasm, causing dense fibrosis; it is considered a hallmark of invasion and malignancy.

Differentiated: A term used to describe how much or how little tumor tissue looks like the normal tissue it came from. Well-differentiated cancer cells look more like normal cells and tend to grow and spread more slowly than poorly differentiated or undifferentiated cancer cells (see Undifferentiated).

Diploidy: The presence of a normal number (two sets) of chromosomes in a cell (see Aneuploidy).

Dysplasia: The overgrowth of immature cells at the location where the number of mature cells is decreasing. This term is used particularly for when cellular abnormality is restricted to new tissues.

Endoscopy: A thin, tube-like instrument with a light and a lens for checking for abnormal areas inside the body.

FISH: Fluorescence *in situ* hybridization for determining the positions of particular genes, for identifying chromosomal abnormalities, and for mapping genes of interest.

Grade, grading: A measure of cell anaplasia (the reversion of differentiation) in a tumor that is based on the resemblance of the tumor to the tissue of origin. Depending on the amount of abnormality, a tumor is graded as 1 (well differentiated; low grade), 2 (moderately differentiated; intermediate grade), 3 (poorly differentiated; high grade), or 4 (undifferentiated; high grade) (see Stage and TNM).

H & E stain: The combined use of hematoxylin (positively charged) and eosin (negative charged) to stain nucleic acids (negatively charged) in blue and amino groups in proteins (negatively charged) in pink, respectively (see IHC).

Hamartoma: A benign, tumor-like, focal malformation resulting possibly from a developmental error. Composed of an abnormal or disorganized mixture of cells and tissues, a hamartoma grows at the same rate as the surrounding tissues and rarely invades or compresses nearby structures significantly. In contrast, a true benign tumor may grow faster than surrounding tissues and compresses nearby structures. Despite its benign histology, a hamartoma may be implicated in some rare, but life-threatening clinical issues such as those associated with neurofibromatosis type 1 and tuberous sclerosis.

A non-neoplastic mass (e.g., hemangioma) can also arise in this way, contributing to misdiagnosis (see Benign tumor).

Hematopoiesis, hemapoiesis: A process in which a hematopoietic stem cell differentiates into a myeloid stem cell, then various blasts, and finally myeloid lineage cells (i.e., red blood cells [erythrocytes], white blood cells [eosinophils, basophils, neutrophils, mast cells, and monocytes but not lymphocytes) and platelets) (see Lymphopoiesis).

HSCT, hematopoietic stem cell transplantation: The intravenous infusion of allogeneic (genetically dissimilar and immunologically incompatible cells from individuals of the same species) or autologous (cells from the same individual) stem cells to reestablish hematopoietic function in patients with damaged or defective bone marrow or immune system. HSCT is often used to treat leukemia, lymphoma, myeloma, anemia, neuroblastoma, germ cell tumors, autoimmune disorders, and other diseases.

Hyperplasia: A disease associated with an increase in the number of normal-looking cells, leading to an enlarged organ; it is also called hypergenesis.

IHC, immunohistochemistry: A technique that exploits the principle of antibodies binding specifically to antigens in biological tissues to visualize the distribution and localization of specific cellular components within cells and in the proper tissue context. Similar to H & E stain, IHC helps detect cellular abnormalities and verifies whether tumor/cancer cells are present at the edge of the material removed (positive margins), are not (negative, not involved, clear, or free margins), or are neither negative nor positive (close margins).

Ki-67: A nuclear protein (also known as KI-67 or MKI67) associated with cellular proliferation and ribosomal RNA transcription. Ki-67 protein is present during all active phases of the cell cycle (G1 [pre-DNA synthesis], S [DNA synthesis], G2 [post-synthesis], and M [mitosis]) but absent in the resting phase (G_0). The fraction of Ki-67-positive tumor cells detected by MIB-1 (the Ki-67 labeling index or MIB-1 labeling index) often correlates to the aggressiveness and thus the clinical course of cancer (see MIB-1).

Lesion: A term in medicine to describe all the abnormal biological tissue changes, such as a cut, a burn, a wound, or a tumor. In cancer, lesion is used interchangeably with tumor, cancer, or neoplasm (see Cancer, Tumor, and Neoplasm).

LOH: Loss of heterozygosity is a gross chromosomal event that results in loss of the entire gene and the surrounding chromosomal region.

Lymphopoiesis, lymphocytopoiesis, or lymphoid hematopoiesis: A process in which a hematopoietic stem cell develops into a lymphoid stem cell, then B- and T-lymphoblasts as well as a natural killer cell precursor, and finally into lymphoid lineage cells (i.e., T- and B-lymphocytes as well as natural killer cells) (see Hematopoiesis).

Malignancy: The state or presence of a malignant tumor.

Malignant tumor: A tumor with the capability of invading surrounding tissues, producing metastases, and recurring after attempted removal.

Metaplasia: The reversible replacement of one differentiated cell type with another mature differentiated cell type.

MIB-1: A monoclonal antibody raised against Ki-67 protein that allows accurate immunohistochemical detection of active or growing cells (see Ki-67).

Mitotic figure, mitosis: Microscopic detection of the chromosomes as tangled, dark-staining threads in cells undergoing mitosis; it is often expressed as mitotic figures per 10 high-power fields (hpf, usually 400-fold magnification; mitotic activity index) or per 1000 tumor cells (mitotic index). As mitotic cell count per 10 hpf equals an area 0.183 mm^2, the American Joint Committee on Cancer (AJCC) specifies that the mitotic rate (the proportion of cells in a tissue undergoing mitosis) be reported as mitoses per mm^2, with a conversion factor of 1 mm^2 equaling 4 hpf.

MRI: Magnetic resonance imaging (also called nuclear magnetic resonance imaging or NMRI) relies on a magnet, radio waves, and a computer to take a series of detailed pictures of affected areas inside the body that help pinpoint the location and dimension of tumor mass, if present. An MRI has a better image resolution than a CT. It includes T1-weighted, T2 weighted, fluid attenuated inversion recovery (FLAIR, also called dark fluid technique), and diffusion weighted imaging (DWI) sequences. T1-weighted images reveal anatomical details and information about venous sinus permeability or pathologic blush (e.g., water and cerebrospinal fluid appears dark; fat and calcification appear white/gray). Use of intravenous contrast gadolinium in T1-weighted sequences further enhances and improves the quality of the images. T2 weighted images provide information about edema, arteries, and sinus permeability (e.g., water appears white/hyperintense; fat and calcification appear gray/dark). FLAIR sequences remove the effects of fluid (which normally covers a lesion) from the resulting images (e.g., cerebrospinal fluid appears dark; edema appears enhanced).

DWI sequences help visualize acute infarction and other inflammatory lesions.

Mutation: A change in the structure of a gene caused by the alteration of single base units in DNA or the deletion, insertion, or rearrangement of larger sections of genes or chromosomes, leading to the formation of a variant that may be transmitted to subsequent generations.

Necrosis: A form of cell injury leading to the premature/unprogrammed death of cells and living tissue caused by autolysis (due to too little blood flowing to the tissue; see Apoptosis).

Neoplasia: A term that describes abnormal growth/proliferation of cells, resulting in a tumor that can be cancerous.

Neoplasm: A new and abnormal growth of tissue in a part of the body; this term is used interchangeably with tumor or cancer.

Oncogene: A gene whose mutation or abnormally high expression can transform a normal cell into a tumor cell.

Parenchyma: The functional tissue of an organ as distinguished from the connective and supporting tissue (see Stroma).

PCR, polymerase chain reaction: A procedure for rapid, *in vitro* production of multiple copies of particular DNA sequences relevant to diagnosis.

PET: A positron emission tomography scan combines a computer-based scan with a radioactive glucose (sugar) injected into a vein to generate a rotating picture of an affected area, with malignant tumor cells showing up brighter due to their more active taking up of glucose than normal cells.

Pleomorphism: A term used in histology and cytopathology to describe variability in the size, shape, and staining of cells and/or their nuclei; it is a feature characteristic of malignant neoplasms and dysplasia.

Radiography: A term used to collectively describe electromagnetic radiation (especially X-ray)-based procedures to visualize the internal structure of a nonuniformly composed and opaque object such as the human body (see MRI and CT).

Radiotherapy: Also called radiation, radiation therapy, or X-ray therapy, involving delivery of radiation externally through the skin or internally (brachytherapy) for destruction of cancer cells or inhibition of their growth.

Stage, staging: As a measurement of the extent to which a tumor has spread, stage is commonly expressed in two ways: clinical and pathological stages. A clinical stage (ranging from 0, I, II, III, to IV) is an estimate of the extent of the tumor after physical exam, imaging studies (e.g., x-rays, CT, MRI) and tumor biopsies as well as other tests (e.g., blood tests), and provides a vital means for deciding the best treatment to use and for comparing the tumor response to treatment. A pathological stage relies on the results of the exam and tests mentioned above, in addition to information obtained during surgery (e.g., the actual degree of tumor spread, nearby lymph node involvement), and thus offers a more precise guide than clinical stage in helping predict treatment response and outcome/prognosis (see Grade and TNM).

Stroma: The parts of a tissue or organ that have a connective and structural role and that do not conduct the specific functions of the organ (e.g., connective tissue, blood vessels, nerves, and ducts) (see Parenchyma).

TNM: A system (i.e., the TNM system) designed by American Joint Committee on Cancer (AJCC) to pathologically stage a solid tumor. T (tumor; TX, T0, T1, T2, T3, and T4) indicates the depth of the tumor invasion—the higher the number, the further the cancer has invaded; N (nodes; NX, N0, N1, N2, and N3) indicates whether the lymph nodes are affected, and how much the tumor has spread to lymph nodes near the original site; and M (metastasis; MX, M0, and M1) indicates whether the tumor has spread to other parts of the body. Thus, the pathological stage of a given tumor may be designated as T1N0MX or T3N1M0 (with numbers after each letter providing further details about the tumor). Knowing a tumor's pathological stage helps in the selection of the most appropriate treatments and gives a more accurate guidance on its prognosis (see Grade and Stage).

Translocation: A segment from one chromosome is transferred to a nonhomologous chromosome (interchromosomal transloaction) or to a new site on the same chromosome (intrachromosomal translocation). A reciprocal translocation occurs when two non-homologous chromosomes swap parts; whereas a non-reciprocal translocation occurs when the transfer of chromosomal material is one way, i.e., another segment does not exchange places with the first segment.

Tumor: A swelling of a part of the body, generally without inflammation, caused by an abnormal growth of tissue, either benign or malignant (see Cancer, Neoplasm, and Lesion).

Tumor suppressor gene: A gene (also known as an antioncogene) that regulates cell division, repairs DNA mistakes, or instructs cells when to die. When a tumor suppressor gene is mutated, cell growth gets out of control (see Apoptosis).

Ultrasound: A device for delivering sound waves that bounce off tissues inside the body like an echo and that records the echoes to create a picture (sonogram) of areas inside the body.

Undifferentiated: The presence of very immature and primitive cells that do not look like cells in the tissue of their origin. Undifferentiated cells are said to be anaplastic,and an undifferentiated cancer is more malignant than a cancer of thetype which is well-differentiated (see Anaplasia and Differentiated).

Index

J

L

M

N

O

P

R

S

W

X

Y

Z

POCKET GUIDES TO
BIOMEDICAL SCIENCES

Series Editor
Dongyou Liu

A Guide to AIDS
Omar Bagasra and Donald Gene Pace

Tumors and Cancers: Brain – Central Nervous System
Dongyou Liu

Tumors and Cancers: Head – Neck – Heart – Lung – Gut
Dongyou Liu

A Guide to Bioethics
Emmanuel A. Kornyo

Tumors and Cancers: Skin – Soft Tissue – Bone – Urogenitals
Dongyou Liu